T0293208

The Rise of Machines
Future of Work in the Age of AI

Adrian David Cheok
Nanjing University of Information Science and Technology
Nanjing, China

Chamari Edirisinghe
Department of Technology, University of Kelaniya
Kelaniya, Sri Lanka

Mangesh Lal Shrestha
Frost & Sullivan, USA
Frost Digital Ventures, USA

CRC Press
Taylor & Francis Group
Boca Raton London New York

CRC Press is an imprint of the
Taylor & Francis Group, an **informa** business

A SCIENCE PUBLISHERS BOOK

First edition published 2025
by CRC Press
2385 NW Executive Center Drive, Suite 320, Boca Raton FL 33431

and by CRC Press
4 Park Square, Milton Park, Abingdon, Oxon, OX14 4RN

CRC Press is an imprint of Taylor & Francis Group, LLC

Library of Congress Cataloging-in-Publication Data (applied for)

ISBN: 978-1-032-58220-7 (hbk)
ISBN: 978-1-032-58221-4 (pbk)
ISBN: 978-1-003-44907-2 (ebk)

DOI: 10.1201/9781003449072

Typeset in Times New Roman
by Prime Publishing Services

Dedicated to Nikola Tesla. Dedicated to the strength and resilience of people who never gave up in the face of change throughout history. With love to Angelina.

Foreword

It is with great admiration that I introduce Professor Adrian David Cheok's latest book [The Rise of the Machines: The Future of Work in the Age of AI]. In this work, Adrian, a distinguished figure in the realm of technology and innovation, presents a lucid and captivating exploration of Artificial Intelligence. This book reflects not just his depth of knowledge, but also his unique ability to make complex ideas accessible and engaging.

One of the most remarkable aspects of this book is its approachability. Adrian has crafted a narrative that is not only insightful for experts but also remarkably accessible for those who may not be deeply entrenched in technological fields. He has a knack for distilling intricate concepts into understandable and relatable content. This makes [The Rise of the Machines: The Future of Work in the Age of AI] a rare gem – a book about AI that speaks to everyone, from curious novices to seasoned professionals.

I have always been fascinated by the potential and challenges that AI presents. This book is a valuable resource that broadens our understanding of AI, encouraging readers to consider its role in our future. Adrian's ability to foresee and articulate the societal and ethical dimensions of AI is truly commendable.

In closing, I applaud Professor Cheok for this exceptional addition to the literature on Artificial Intelligence. His work is an invitation for all of us to engage with the subject in a thoughtful and imaginative manner.

Yair Goldfinger
Co-founder and CEO, AppCard
Co-founder, Dotomi
Co-founder, ICQ
December 2023   New York, USA

Preface

As an expert in both Artificial Intelligence and Virtual Reality, it is my honor to introduce Professor Adrian David Cheok's seminal work [The Rise of the Machines: The Future of Work in the Age of AI]. This book, a profound exploration of the intersection between AI and the human experience, resonates deeply with the core of my own research and academic pursuits at the Nanjing University of Information Science and Technology.

Adrian's book is a beacon of knowledge in a sea of technological change. His approach to AI is not only technically sound but also deeply humane, reflecting a nuanced understanding of how AI reshapes our world. As we delve into this new age, where virtual reality and AI are not just tools but partners in our daily lives, this book provides critical insights and guidance.

Throughout my career, I have seen few works that so eloquently bridge the gap between advanced technological concepts and their practical implications. Adrian's expertise shines through in every page, making this book an indispensable resource for anyone interested in the future of AI, whether they are students, professionals, or simply curious minds.

In conclusion [The Rise of the Machines: The Future of Work in the Age of AI] is more than just a book; it's a roadmap for navigating the challenges and opportunities of our times. I commend Professor Cheok for his outstanding contribution to our understanding of AI and its impact on society.

Professor Zhigeng Pan
Dean, School of Artificial Intelligence (Future Technology)
Nanjing University of Information Science and Technology
Editor-in-Chief, Metaverse
Director, Research Center for Metaverse Engineering in Jiangsu
December 2023   Province Nanjing, China

Contents

Chapter 1

The Evolving Landscape of Artificial Intelligence: Insights and Narratives

1.1 Introduction

1.1.1 Overview of the AI Landscape

Artificial intelligence (AI) has emerged as one of the most transformative technological paradigms of the 21st century. Its evolution from theoretical constructs to practical, impactful applications represents a significant leap in both computational capabilities and societal impact. The landscape of AI is characterized by rapid advancements, diverse applications, and a complex interplay of ethical, economic, and technological considerations.

The genesis of AI can be traced back to the mid-20th century, with foundational work by pioneers such as Alan Turing and John McCarthy. However, it is in the recent decades that AI has truly flourished, propelled by advancements in machine learning, neural networks, and big data analytics. This period has witnessed the emergence of AI systems capable of surpassing human performance in specific tasks, ranging from playing strategic games like Go to diagnosing diseases with higher accuracy than human experts.

Central to the AI narrative are the stories of individuals and organizations that have played pivotal roles in shaping its trajectory. Figures like Elon Musk and Larry Page, and entities such as Google's DeepMind and OpenAI, have not only contributed to technological breakthroughs but have also influenced the public discourse on the potential and perils of AI. Their personal journeys, marked by both collaboration and contention, embody the dynamism and complexity of the AI field.

1.1.2 Purpose and Scope of the Chapter

This chapter aims to provide an in-depth exploration of the modern AI landscape, with a focus on the key figures, institutions, and events that have shaped its development. Drawing upon the insightful reporting of Metz, Weise, Grant, and Isaac in their New York Times article "Ego, Fear and Money How the AI Fuse Was Lit" [1], this chapter will delve into the intriguing narratives and pivotal moments that define the current state of AI.

The scope of this chapter extends beyond a mere chronological account of AI's evolution. It seeks to weave the technical advancements with the personal stories of the industry's luminaries, the ethical dilemmas they grapple with, and the societal implications of their work. By doing so, the chapter aims to present a nuanced view of AI that captures both its scientific brilliance and the human elements that drive its progress.

In the following sections, we will explore the early days of AI, the rise and impact of influential AI entities like DeepMind and OpenAI, the ethical and philosophical debates surrounding AI, and the future directions of this rapidly evolving field. Through this comprehensive examination, the chapter will illuminate the multifaceted nature of AI and its profound implications for the future of humanity.

1.2 Historical Context of AI Development

1.2.1 Early Days of Artificial Intelligence

The odyssey of artificial intelligence (AI) is a saga of human ingenuity and relentless pursuit of knowledge. Tracing its origins back to the mid-20th century, AI began as a confluence of ideas from mathematics, logic, and computer science. The term "artificial intelligence" itself was coined in 1956 during the Dartmouth Conference, a gathering that marked the official birth of the field [2].

This era was marked by visionary thinkers like Alan Turing, whose seminal paper on the "imitation game" laid the groundwork for what would become the Turing Test, a benchmark for machine intelligence [3]. Turing's insights, along with the mathematical foundations laid by others such as Kurt Gödel and Claude Shannon, set the stage for the first wave of AI research.

1.2.2   Key Milestones in AI Research

The evolution of AI has been punctuated by several key milestones that have significantly advanced the field. The development of neural networks, a concept inspired by the human brain, has been a cornerstone of AI research. This was complemented by the emergence of deep learning in the early 2000s, which revolutionized the field by enabling AI systems to learn and make decisions with unprecedented accuracy [4].

In the 1960s, the development of rule-based expert systems, such as MYCIN, marked a significant advancement in AI's application in domains like medicine [5]. The 1980s witnessed a paradigm shift with the introduction of machine learning algorithms, which moved AI from rule-based systems to models capable of learning from data [6].

The late 20th and early 21st centuries saw AI breakthroughs in natural language processing and computer vision, leading to the development of systems like IBM's Watson and Google's AlphaGo. These systems demonstrated AI's ability to not only analyze vast amounts of data but also to outperform humans in complex tasks like playing the game of Go [7].

As we continue to explore the potential of AI, it is crucial to acknowledge the contributions of these milestones and the researchers behind them. Their work has not only advanced the field technically but has also sparked important discussions about the ethical and societal implications of AI.

1.3   DeepMind: Inception and Acquisition

1.3.1   Founding of DeepMind

DeepMind Technologies, a trailblazer in the realm of artificial intelligence, was founded in September 2010 by Demis Hassabis, Shane Legg, and Mustafa Suleyman. The inception of DeepMind was driven by a profound vision: to solve intelligence and then use it to solve everything else [8].

Hassabis, a former child prodigy in chess, had a deep background in neuroscience and computer science. His vision for DeepMind was to create a general-purpose learning algorithm, one that could learn from raw data and perform a wide variety of tasks, much like the human brain [8]. This ambitious goal set DeepMind apart from other AI ventures, which were primarily focused on specific applications of AI.

The early days of DeepMind were marked by a blend of academic research and a startup culture. The founders brought together experts in machine learning, neuroscience, and deep learning, fostering an environment where cutting-edge research and innovative thinking thrived [9].

1.3.2 Google's Acquisition and Its Conditions

The acquisition of DeepMind by Google in January 2014 for approximately $500 million was a watershed moment in AI industry [10]. This acquisition was not just a business transaction; it was a strategic move by Google to position itself at the forefront of the AI revolution.

One of the most intriguing aspects of this acquisition was the conditions put forth by DeepMind's founders. Concerned about the ethical implications of AI, they stipulated that Google establish an AI ethics board to oversee the responsible development of artificial intelligence [11]. This condition highlighted the founders' commitment to the safe and ethical use of AI, a concern that has become increasingly relevant in the broader conversation about AI's role in society.

The acquisition also included agreements on the autonomy of DeepMind. The founders were adamant that DeepMind should continue its research-focused mission, relatively unhindered by the commercial interests of its parent company. This autonomy has allowed DeepMind to continue pushing the boundaries of AI research, leading to groundbreaking achievements such as the development of AlphaGo, the first computer program to defeat a human professional Go player [12].

In conclusion, the story of DeepMind's founding and its acquisition by Google is a testament to the transformative power of vision, ambition, and ethical consideration in the field of artificial intelligence. It serves as a compelling narrative of how a small startup with big dreams can redefine the landscape of technology and spark a global conversation about the future of AI.

1.3.3 Formation of OpenAI

OpenAI, an artificial intelligence research laboratory, was founded in December 2015 as a response to the growing concerns about the potential monopolization of AI technology by large corporations and the ethical implications of AI development. The organization was established by a group of high-profile individuals from the tech industry, including Elon Musk, Sam Altman, Greg Brockman, Ilya Sutskever, Wojciech Zaremba, and John Schulman, among others [13].

The founding of OpenAI was driven by a mission to ensure that artificial general intelligence (AGI) benefits all of humanity. The founders believed that by making their research open and accessible, they could democratize AI technology, thereby preventing any single entity from gaining overwhelming control over it [14]. This was a significant departure from the traditional model of AI research, which was often shrouded in secrecy and confined within the walls of corporate labs.

OpenAI's approach to AI research was grounded in the belief that collaboration and transparency were key to developing safe and beneficial AI. The organization initially operated as a non-profit entity, allowing it to focus on long-term research objectives without the pressure of generating immediate financial returns. This model attracted some of the brightest minds in AI research, who were drawn to OpenAI's ethos of open collaboration and its commitment to addressing some of the most challenging problems in AI [15].

One of the early focuses of OpenAI was on reinforcement learning, a type of machine learning technique where AI systems learn to make decisions by interacting with their environment. This approach was seen as a crucial step towards building AGI, as it mimicked the way humans learn from experience [16]. OpenAI's research in this area led to several notable achievements, including the development of AI systems that could achieve superhuman performance in a range of complex tasks, from playing video games to controlling robotic hands.

The formation of OpenAI marked a pivotal moment in the AI landscape. It represented a collective effort by some of the industry's leading figures to steer the development of AI towards a future that was safe, ethical, and beneficial for all. As OpenAI continues to push the boundaries of AI research, its founding principles remain a guiding light, reminding us of the power of open collaboration and the importance of aligning technological advancements with human values.

1.3.4 Development and Impact of GPT-3

The development of GPT-3 (Generative Pre-trained Transformer 3) by OpenAI represents a monumental leap in the field of natural language processing (NLP) and artificial intelligence. GPT-3, unveiled in June 2020, is the third iteration of the Generative Pre-trained Transformer series and is by far the most powerful and sophisticated [17].

GPT-3's architecture is a marvel of engineering and machine learning design. It is a language model with 175 billion parameters, making it one of the largest and most complex AI models ever created [18]. This immense scale allows GPT-3 to generate human-like text with unprecedented accuracy, making it capable of a wide range of tasks, from writing essays to composing poetry, and even coding.

The development of GPT-3 was rooted in the advancements of its predecessors, GPT and GPT-2. However, GPT-3's scale and capabilities far exceed those of earlier models. Its ability to generate coherent and contextually relevant text across a wide range of subjects without task-specific training is a significant advancement in NLP [19].

GPT-3's impact extends beyond the realm of academic research into practical applications. Its versatility has sparked a wave of innovations, with companies and developers exploring its use in various fields such as content creation, customer service, and even software development [20]. The model's ability to understand and generate human-like text has opened new possibilities for human-computer interaction, making AI more accessible and useful for a broader range of applications.

However, the development of GPT-3 has also raised important questions about the ethical and societal implications of powerful language models. Concerns about bias, misuse, and the broader impact on the job market have been at the forefront of discussions surrounding GPT-3 [21]. OpenAI has acknowledged these challenges, emphasizing the importance of responsible development and deployment of AI technologies.

In conclusion, GPT-3 stands as a testament to the rapid advancements in AI and NLP. Its development not only marks a significant milestone in the journey of OpenAI but also sets a new standard for what is possible in the realm of artificial intelligence. As we continue to explore the capabilities and applications of models like GPT-3, it is crucial to navigate the ethical landscape with as much rigor as we apply to the technological one, ensuring that the benefits of such technologies are shared equitably across society.

1.3.5 Competitive Dynamics in AI Development

The race to dominate the field of artificial intelligence (AI) in Silicon Valley has been marked by intense competition, strategic alliances, and significant investments. This race is not just about technological supremacy but also about shaping the future of how AI will impact society, economy, and everyday life [22].

1.3.5.1 The Early Pioneers and Their Vision

The early days of AI development in Silicon Valley were characterized by visionary entrepreneurs and technologists who foresaw the immense potential of AI Companies like Google, Apple, Facebook, and Amazon, among others, began investing heavily in AI research and development, recognizing its potential to revolutionize various industries [23].

Google's acquisition of DeepMind in 2014 for $500 million was a significant milestone in Silicon Valley's AI race. This acquisition not only demonstrated Google's commitment to leading in AI but also set off a wave of investments and acquisitions by other tech giants [24]. Apple's acquisition of Turi and Amazon's acquisition of Orbeus were part of this trend, as major players sought to bolster their AI capabilities.

1.3.5.2 The Rise of Startups and Innovation Culture

Silicon Valley's AI race was further fueled by the emergence of startups specializing in various aspects of AI, such as machine learning, deep learning, and neural networks. These startups, often founded by leading researchers and experts in AI, brought fresh perspectives and innovative approaches to the field [25].

The culture of innovation in Silicon Valley, characterized by risk-taking and a relentless pursuit of breakthroughs, played a crucial role in driving AI advancements. The region's unique ecosystem, comprising universities, research institutions, and a vibrant venture capital scene, provided the perfect breeding ground for AI innovation [26].

1.3.5.3 Collaborations and Talent Wars

As the AI race intensified, collaborations between tech giants and academic institutions became increasingly common. These collaborations aimed to combine the research prowess of academia with the practical application and scaling capabilities of the industry [27].

The competition for talent in AI also became a critical aspect of the race. Silicon Valley companies aggressively recruited top talent from universities and competitors, leading to a talent war that drove up salaries and created a highly competitive job market for AI experts [28].

1.3.5.4 Ethical Considerations and Societal Impact

The rapid advancements in AI and the fierce competition in Silicon Valley raised important questions about the ethical implications of AI Concerns about privacy, bias, job displacement, and the broader societal impact of AI became central topics of discussion [29].

Tech companies and startups began to recognize the importance of addressing these ethical concerns. Initiatives like the Partnership on AI, which include Amazon, Google, Facebook, IBM, Microsoft, and Apple, were established to promote ethical practices in AI development and to ensure that AI technologies are used for the benefit of humanity [30].

1.3.5.5 Looking to the Future

As we look to the future, the competitive dynamics in AI development in Silicon Valley continue to evolve. The race is no longer just about who can develop the most advanced AI technologies but also about who can responsibly guide the development of these technologies in a way that benefits society as a whole [31].

The ongoing AI race in Silicon Valley is a testament to the region's enduring status as a global hub of innovation and technological advancement. It is a race that is not only shaping the future of technology but also the future of how we live, work, and interact with the world around us.

1.3.6 Major Players and Their Strategies

The landscape of artificial intelligence in Silicon Valley is dominated by a few key players, each with its unique strategy and approach to AI development. These major players include tech giants like Google, Apple, Facebook, Amazon, and

Microsoft, as well as influential startups and research organizations like OpenAI and DeepMind [22].

1.3.6.1 Google: An Early Advocate for Deep Learning

Google has been a pioneer in AI, particularly in the field of deep learning. The company's strategy has been to integrate AI into all its products and services, making AI a core component of its business model [23]. Google's acquisition of DeepMind in 2014 significantly bolstered its AI capabilities, leading to breakthroughs like AlphaGo [24]. Google has also focused on developing TensorFlow, an open-source AI software library, to democratize access to machine learning tools [32].

1.3.6.2 Apple: Focusing on Consumer Privacy and On-Device AI

Apple's approach to AI has been markedly different, with a strong emphasis on consumer privacy. The company has invested in developing AI capabilities that run on the device itself, rather than relying on cloud-based systems. This strategy not only addresses privacy concerns but also improves the speed and efficiency of AI-powered features on Apple devices [25].

1.3.6.3 Facebook: Advancing AI for Social Connectivity

Facebook has focused its AI efforts on enhancing social connectivity and interaction. The company has developed advanced machine learning models to personalize content, target advertisements, and moderate content. Facebook AI Research (FAIR) has been instrumental in advancing the field of AI through research in areas like natural language processing and computer vision [33].

1.3.6.4 Amazon: Pioneering in AI-Powered Commerce

Amazon's AI strategy revolves around leveraging machine learning to revolutionize e-commerce and cloud computing. The company's AI assistant, Alexa, is a prime example of how Amazon is integrating AI into consumer products. Additionally, Amazon Web Services (AWS) offers a range of AI and machine learning services to businesses, driving innovation in cloud-based AI solutions [34].

1.3.6.5 Microsoft: Balancing Commercial and Ethical AI

Microsoft has taken a balanced approach to AI, focusing on both commercial applications and ethical considerations. The company has integrated AI into its suite of products while also investing in research to address issues like AI fairness, reliability, and safety. Microsoft's partnership with OpenAI is a strategic move to stay at the forefront of AI advancements [30].

1.3.6.6   OpenAI and DeepMind: Research-Oriented Approaches

OpenAI and DeepMind, although not traditional tech giants, have significantly influenced the AI landscape. OpenAI's development of GPT-3 has set new standards in natural language processing, while DeepMind's work in deep reinforcement learning has led to groundbreaking achievements in AI Both organizations focus heavily on research and have a commitment to ethical AI development [17, 8].

In conclusion, the strategies of these major players in Silicon Valley's AI race are as diverse as they are impactful. From integrating AI into consumer products to advancing the frontiers of AI research, each player contributes uniquely to the evolution of AI. As the race continues, their collective efforts are shaping not only the future of technology but also the societal and ethical landscape in which AI operates.

1.3.6.7   Influential Figures in Silicon Valley's AI Race

The AI race in Silicon Valley has been significantly shaped by influential figures whose visions, decisions, and actions have had a profound impact on the direction of AI development.

Elon Musk: A Proponent of Ethical AI

Elon Musk, the CEO of Tesla and SpaceX, has been a vocal advocate for the ethical development and regulation of AI Musk's concerns about the potential risks of AI led him to co-found OpenAI in 2015, with the goal of ensuring that AI benefits all of humanity [35]. His stance on AI has been both cautionary and proactive, often emphasizing the need for safeguards against the misuse of AI technologies [36].

Demis Hassabis: The Mind Behind DeepMind

Demis Hassabis, co-founder of DeepMind, stands out as a pivotal figure in the AI landscape. A former chess prodigy and neuroscientist, Hassabis has been instrumental in advancing the field of deep learning and reinforcement learning. Under his leadership, DeepMind achieved a significant milestone with AlphaGo, which defeated the world champion in the game of Go [8]. Hassabis's vision for DeepMind extends beyond gaming; he aims to solve complex scientific and global challenges through AI [24].

Sam Altman: Leading OpenAI's Ambitious Quest

Sam Altman, the CEO of OpenAI, has played a crucial role in shaping the organization's direction and strategy. Altman's leadership has been key in OpenAI's transition from a non-profit to a capped-profit model, a move designed to balance the need for funding with the organization's ethos of widely sharing AI benefits

[37]. Under his guidance, OpenAI has made significant strides in language processing with GPT-3, pushing the boundaries of what AI can achieve [17].

Jeff Dean: Google's AI Visionary

Jeff Dean, Senior Vice President of Google Research and Health, has been a driving force behind Google's AI initiatives. Dean's work on large-scale deep learning systems has been foundational to Google's AI advancements. His contributions to TensorFlow and Google Brain have democratized access to machine learning tools, enabling a wide range of applications across various sectors [10].

Yann LeCun: Pioneering Deep Learning at Facebook

Yann LeCun, Chief AI Scientist at Facebook, is renowned for his work in deep learning and convolutional neural networks. LeCun's research has been integral to the development of AI technologies used in Facebook's platforms. His commitment to advancing AI research while addressing ethical concerns has been a hallmark of his tenure at Facebook [38].

These individuals, among others, have been key players in Silicon Valley's AI race. Their contributions, ranging from groundbreaking research to ethical advocacy, have not only propelled technological advancements but also shaped the conversation around the future of AI and its role in society.

1.3.7   Debates Over AI Ethics

The rapid advancement of artificial intelligence has sparked a wide range of ethical debates, addressing concerns about the impact of AI on society, the economy, and individual rights. These debates encompass a variety of issues, including privacy, bias, accountability, and the future of employment.

1.3.7.1   Privacy and Surveillance

One of the most pressing ethical concerns in the era of AI is privacy. With the increasing capability of AI systems to process and analyze large datasets, there is a growing worry about the potential for mass surveillance and the erosion of personal privacy [39]. The use of AI in facial recognition technology, particularly by law enforcement and government agencies, has raised significant concerns about civil liberties and the potential for abuse [40].

1.3.7.2   Bias and Fairness

Another critical area of ethical concern is the potential for bias in AI systems. There is a growing body of evidence that AI algorithms, particularly in areas like facial recognition and decision-making systems, can perpetuate and amplify societal biases [41]. This has led to calls for more rigorous testing and auditing of AI systems to ensure fairness and prevent discrimination [42].

1.3.7.3 Accountability and Transparency

The issue of accountability in AI decision-making is another area of ethical debate. As AI systems become more complex, determining who is responsible for the decisions made by these systems becomes increasingly challenging [29]. This has led to discussions about the need for greater transparency in AI algorithms and the development of frameworks to ensure accountability in AI decision-making [43].

1.3.7.4 Impact on Employment

The impact of AI on employment is a topic of significant debate. While some argue that AI will lead to job displacement and increased inequality, others believe that AI will create new job opportunities and increase productivity [44]. This debate revolves around the need for policies to manage the transition and ensure that the benefits of AI are broadly shared [45].

1.3.7.5 The Way Forward

Addressing these ethical concerns requires a collaborative effort among technologists, policymakers, and other stakeholders. Initiatives like the Montreal Declaration for Responsible Development of Artificial Intelligence and the IEEE Global Initiative on Ethics of Autonomous and Intelligent Systems are examples of efforts to create ethical guidelines for AI development [11].

In conclusion, the ethical considerations in AI are complex and multifaceted. As AI continues to evolve, it is imperative that these ethical debates are at the forefront of discussions about the future of AI development. Only through careful consideration and proactive management of these ethical challenges can we ensure that AI serves the greater good of society.

1.3.8   Role of Ethics Boards and Oversight

The rapid advancement and integration of artificial intelligence (AI) into various sectors have necessitated the establishment of ethics boards and oversight committees. These entities play a crucial role in guiding the ethical development and deployment of AI technologies, ensuring that they align with societal values and norms.

1.3.8.1 The Need for Ethical Oversight in AI

The development of AI technologies raises complex ethical questions that go beyond traditional boundaries. Issues such as data privacy, algorithmic bias, and the potential for misuse necessitate a structured approach to ethical decision-making [43]. Ethics boards and oversight committees are tasked with addressing these challenges, providing guidance and recommendations to ensure responsible AI development.

1.3.8.2 Composition and Function of Ethics Boards

Ethics boards in the context of AI are typically composed of experts from diverse fields, including technology, law, ethics, sociology, and philosophy. This multidisciplinary approach is crucial for addressing the multifaceted nature of AI ethics [11]. These boards function by setting guidelines, reviewing AI projects and initiatives, and advising on best practices. They also play a role in fostering public dialogue and understanding of AI ethics.

1.3.8.3 Case Studies: Google's AI Ethics Board and OpenAI's Charter

Google's Advanced Technology External Advisory Council (ATEAC), although short-lived, was an example of an attempt to establish an external ethics board to oversee its AI initiatives [10]. The council was intended to provide recommendations on complex issues like facial recognition and fairness in machine learning. Similarly, OpenAI's Charter outlines its commitment to developing AI in a safe and beneficial manner, with oversight mechanisms to ensure adherence to its principles [37].

1.3.8.4 Challenges and Criticisms

Ethics boards in AI face several challenges, including questions about their effectiveness, representativeness, and independence. Critics argue that these boards, often set up by the companies developing AI, may lack the necessary autonomy to provide unbiased oversight [46]. Additionally, the rapid pace of AI development can make it difficult for ethics boards to keep up with the latest advancements and implications.

1.3.8.5 The Future of AI Ethics Oversight

Looking forward, the role of ethics boards and oversight in AI will likely continue to evolve. There is a growing recognition of the need for more robust and independent forms of oversight, potentially involving governmental and international bodies [47]. As AI continues to advance, the development of comprehensive and enforceable ethical frameworks will be crucial for harnessing the benefits of AI while mitigating its risks.

In conclusion, ethics boards and oversight play a pivotal role in the responsible development of AI technologies. Their guidance is essential for navigating the ethical complexities of AI and ensuring that its development aligns with societal values and norms. As the AI landscape continues to evolve, the importance of effective ethical oversight cannot be overstated.

1.3.9 AI and Job Displacement

The integration of artificial intelligence (AI) into various sectors of the economy has sparked significant debate regarding its impact on employment and job displacement. This complex issue encompasses both the potential for AI to replace human labor in certain tasks and the creation of new job opportunities in emerging fields.

1.3.9.1 The Automation of Jobs

One of the most immediate concerns regarding AI is its capability to automate tasks traditionally performed by humans. This automation potential is particularly pronounced in sectors like manufacturing, transportation, and customer service, where routine and repetitive tasks are prevalent [44]. Studies have shown that AI and robotics could replace a significant portion of existing jobs, leading to substantial shifts in the labor market [48].

1.3.9.2 The Changing Nature of Work

While the automation of jobs is a concern, it is also important to consider how AI is changing the nature of work. AI technologies are not just replacing jobs but also transforming them, requiring new skill sets and roles. For instance, the demand for data scientists, AI specialists, and machine learning engineers is growing, reflecting the need for expertise in developing and managing AI systems [45].

1.3.9.3 Economic Benefits and Productivity Gains

On the positive side, AI has the potential to drive significant economic benefits and productivity gains. AI can improve efficiency, reduce errors, and enable the development of new products and services. This can lead to economic growth and increased competitiveness for businesses that effectively leverage AI technologies [49].

1.3.9.4 Policies for Managing AI Job Displacement

The potential job displacement caused by AI has led to discussions about the need for policies to manage this transition. Proposals include retraining programs for workers displaced by AI, incentives for companies to create new jobs, and even the consideration of universal basic income as a way to mitigate the impact of automation on employment [50].

1.3.9.5   The Long-Term Outlook

In the long term, the impact of AI on jobs may be more nuanced than the initial fears of widespread unemployment. History has shown that technological advancements, while disruptive in the short term, can lead to the creation of new industries and job categories in the long run [51]. The key challenge will be ensuring that the workforce is prepared for these changes and that the benefits of AI are broadly distributed across society.

In conclusion, the economic implications of AI on job displacement are multifaceted and complex. While there are legitimate concerns about the potential for job loss, there are also opportunities for job creation and economic growth. Navigating this transition will require thoughtful policies, investment in education and training, and a commitment to ensuring that the benefits of AI are shared equitably.

1.3.10   Utopian vs. Dystopian Views

The philosophical discourse on artificial intelligence (AI) oscillates between utopian and dystopian perspectives, each offering a distinct vision of the future shaped by AI technologies. These perspectives reflect the diverse and often conflicting opinions on the potential impact of AI on society.

1.3.10.1   Utopian Views: AI as a Force for Good

Utopian views on AI are characterized by optimism about the potential of AI to solve some of humanity's most pressing problems. Proponents of this view argue that AI can drive significant advancements in fields such as healthcare, environmental protection, and education, ultimately leading to a better, more efficient, and equitable world [52].

Advocates of the utopian perspective envision a future where AI augments human capabilities, automates tedious and dangerous tasks, and provides solutions to complex global challenges. They argue that AI has the potential to significantly improve the quality of life, enhance human creativity, and even extend human capabilities beyond current limitations [53].

1.3.10.2   Dystopian Views: AI as a Threat

In contrast, dystopian views of AI are grounded in skepticism and concern about the potential negative consequences of advanced AI systems. Critics in this camp fear that unchecked AI development could lead to loss of privacy, increased surveillance, job displacement, and widening social and economic inequalities [53].

Dystopian scenarios often include fears of AI surpassing human intelligence (a phenomenon known as singularity), leading to scenarios where humans could lose control over AI systems. This perspective raises concerns about the ethical implications of autonomous AI systems, the potential for misuse of AI in warfare and surveillance, and the existential risk that superintelligent AI could pose to humanity [31].

1.3.10.3 Balancing the Perspectives

The debate between utopian and dystopian views on AI is not just philosophical but also practical, influencing policy and research directions in the field of AI. It is increasingly recognized that a balanced approach is needed, one that harnesses the benefits of AI while mitigating its risks [11].

This balanced approach involves proactive engagement with ethical considerations, robust regulatory frameworks, and ongoing dialogue among technologists, policymakers, ethicists, and the broader public. The goal is to steer AI development in a direction that maximizes its positive impact on society while minimizing potential harms [27].

In conclusion, the philosophical perspectives on AI reflect the complex and multifaceted nature of this technology. As AI continues to evolve, it is imperative that these discussions inform the development and deployment of AI systems, ensuring that they serve the broader interests of society and contribute to a future that is beneficial for all.

1.3.10.4 Musk and Page's Divergent Opinions

The discourse on the future of artificial intelligence (AI) is often highlighted by the contrasting views of Elon Musk and Larry Page, two of the most influential figures in the tech industry. Their differing opinions encapsulate the broader debate between optimism and caution in the face of advancing AI technology.

Elon Musk: A Cautionary Stance on AI

Elon Musk, CEO of Tesla and SpaceX, has been vocal about his concerns regarding the unchecked advancement of AI Musk's cautionary stance is rooted in the fear that AI, particularly superintelligent systems, could pose existential risks to humanity if not properly controlled and regulated [35]. He has advocated for proactive regulatory measures to ensure that AI development is aligned with human safety and ethical standards [36]. Musk's views have been influential in shaping public discourse on the potential dangers of AI and the need for ethical oversight.

Larry Page: An Optimistic View of AI

In contrast, Larry Page, co-founder of Google, holds a more optimistic view of AI Page envisions AI as a transformative force that can significantly improve human life and solve complex global challenges [54]. He believes that AI has the potential to drive innovation, efficiency, and economic growth. Under Page's leadership, Google has invested heavily in AI research and development, reflecting his belief in the positive impact of AI on society [24].

The Impact of Their Views on the Tech Industry

The divergent views of Musk and Page on AI have had a significant impact on the tech industry and the broader public's perception of AI. Their opinions have influenced policy discussions, research priorities, and public sentiment towards AI [36]. While Musk's warnings have raised awareness about the potential risks of AI, Page's optimism has highlighted the technology's potential benefits.

Balancing Caution and Optimism

The debate between Musk and Page reflects the broader challenge of balancing caution and optimism in the development of AI. It underscores the need for a nuanced approach that recognizes both the potential benefits and risks of AI. This balance is crucial for guiding the responsible development of AI technologies that can contribute positively to society while mitigating potential harms [31].

In conclusion, the divergent opinions of Elon Musk and Larry Page on AI represent the dual aspects of caution and optimism that characterize the current discourse on AI. Their views serve as a reminder of the complex and multifaceted nature of AI development and the importance of considering a wide range of perspectives in shaping the future of this transformative technology.

1.3.11 Representation of AI in Films and Literature

The portrayal of artificial intelligence (AI) in films and literature has played a significant role in shaping public perception and understanding of AI. These cultural media have explored the myriad possibilities of AI, from utopian visions of AI as a benevolent helper to dystopian depictions of AI as a threat to humanity.

1.3.11.1 AI in Science Fiction Literature

Science fiction literature has long been a fertile ground for exploring the implications of artificial intelligence. Classic works like Isaac Asimov's "I, Robot" series have laid the groundwork for the public's understanding of AI, introducing concepts such as the Three Laws of Robotics [55]. These stories often delve into the ethical and moral dilemmas posed by AI, exploring themes of autonomy, consciousness, and the relationship between humans and machines [56].

1.3.11.2 AI in Film: From '2001: A Space Odyssey' to 'Ex Machina'

In film, AI has been depicted in various ways, often reflecting the societal and technological concerns of the times. Stanley Kubrick's "2001: A Space Odyssey" introduced audiences to HAL 9000, an AI that becomes dangerously self-aware, highlighting fears about losing control over intelligent machines [57]. More recently, films like "Ex Machina" and "Her" have presented more nuanced portrayals of AI, examining the complexities of AI consciousness and the potential for emotional connections between humans and AI [58].

1.3.11.3 The Impact of Media Portrayals on AI Perception

The portrayal of AI in films and literature has a profound impact on how the public perceives AI technologies. While these portrayals can help demystify AI and spark interest in the field, they can also perpetuate misconceptions and fears about AI [59]. The dramatization of AI as either a savior or a destroyer often overlooks the more nuanced and complex reality of AI development.

1.3.11.4 The Role of Media in Shaping AI Discourse

Media representations of AI also play a crucial role in shaping the discourse around AI ethics, policy, and research. By bringing AI into the cultural conversation, films and literature can help foster a broader dialogue about the societal implications of AI and the direction of its development [59].

In conclusion, the representation of AI in films and literature is not just a reflection of societal attitudes towards technology but also a powerful shaper of those attitudes. As AI continues to advance, it is important for these cultural portrayals to evolve as well, providing a balanced and informed perspective on the potential and challenges of AI.

1.3.11.5 Public Perception of AI

The public perception of artificial intelligence (AI) is shaped by a confluence of factors, including media portrayals, personal experiences with technology, and societal discourse. Understanding how the public views AI is crucial for the development and integration of these technologies in society.

Influence of Media on Public Perception

Media representations of AI, whether in news, films, or literature, significantly influence public perception. Dramatic portrayals of AI in science fiction films and books often skew towards extreme scenarios, either utopian or dystopian, which can lead to misconceptions about the capabilities and intentions of AI systems [59]. News coverage that sensationalizes AI advancements or failures also contributes to shaping public opinion [49].

Experiences with Consumer AI Products

Everyday experiences with consumer AI products, such as virtual assistants, recommendation systems, and autonomous vehicles, also inform public perception.

Positive experiences can lead to increased trust and acceptance of AI, while negative experiences or high-profile failures can fuel skepticism and fear [22].

Societal Discourse and Educational Outreach

Societal discourse, including educational outreach and public debates, plays a role in shaping the public's understanding of AI. Initiatives aimed at demystifying AI and providing accurate information about its capabilities and limitations can help foster a more nuanced and informed public perception [27].

Public Concerns and Hopes for AI

Public concerns about AI often revolve around issues of privacy, job displacement, and the ethical use of technology. Conversely, there is also optimism about the potential of AI to improve healthcare, enhance productivity, and solve complex global challenges [31]. Balancing these concerns and hopes is key to achieving societal acceptance of AI.

The Role of Policy and Regulation

Public perception of AI is also influenced by policy and regulation. Effective policies that address privacy, safety, and ethical concerns can increase public trust in AI. Conversely, a lack of regulation or high-profile regulatory failures can lead to increased public skepticism and fear [47].

In conclusion, public perception of AI is multifaceted and evolving. It is influenced by a variety of factors, including media portrayals, personal experiences, societal discourse, and policy. Understanding and addressing public concerns and misconceptions is crucial for the responsible development and integration of AI in society.

1.3.12   Key AI Technologies and Their Evolution

The field of artificial intelligence (AI) has witnessed a series of significant technological advances and breakthroughs over the years. These technologies have evolved from basic machine learning algorithms to sophisticated systems capable of performing complex tasks with a high degree of autonomy and intelligence.

1.3.12.1   From Machine Learning to Deep Learning

The evolution of AI technologies began with the development of machine learning, where algorithms use statistical techniques to enable computers to 'learn' from and make predictions or decisions based on data [60]. The advent of deep learning, a subset of machine learning involving neural networks with multiple layers, marked a significant advancement in the field. Deep learning has enabled breakthroughs in areas such as image and speech recognition, natural language processing, and autonomous vehicles [61].

1.3.12.2 Neural Networks and Their Impact

Neural networks, inspired by the structure and function of the human brain, have been central to the evolution of AI. These networks consist of interconnected nodes (neurons) that process and transmit information. The development of convolutional neural networks (CNNs) and recurrent neural networks (RNNs) has been particularly influential, enabling significant progress in computer vision and sequential data processing, respectively [62].

1.3.12.3 Reinforcement Learning and Game-Playing AI

Reinforcement learning, where AI systems learn to make decisions by receiving rewards or penalties for their actions, has led to the development of AI that can master complex games. Notable examples include DeepMind's AlphaGo, which defeated a world champion Go player, and OpenAI's systems that have excelled in various video games [9].

1.3.12.4 Natural Language Processing and Generative Models

Advancements in natural language processing (NLP) have revolutionized how machines understand and interact with human language. The development of generative models like GPT-3 by OpenAI has further pushed the boundaries, enabling machines to generate human-like text and engage in conversations with unprecedented coherence and relevance [17].

1.3.12.5 Autonomous Systems and Robotics

The integration of AI into robotics has led to the creation of autonomous systems capable of performing tasks without human intervention. This includes everything from self-driving cars to drones and robotic assistants. These systems combine various AI technologies, including computer vision, sensor fusion, and decision-making algorithms, to navigate and interact with the physical world [63].

In conclusion, the evolution of AI technologies has been marked by continuous innovation and breakthroughs. From the early days of machine learning to the current era of deep learning and autonomous systems, AI technology has transformed numerous industries and continues to be a driving force in the advancement of intelligent systems.

1.3.12.6 Notable AI Achievements

The field of artificial intelligence (AI) has seen a multitude of remarkable achievements that have pushed the boundaries of technology and opened new avenues for application and research. These achievements not only demonstrate the capabilities of AI but also highlight the rapid pace of innovation in the field.

Breakthroughs in Deep Learning and Neural Networks

One of the most significant achievements in AI has been the development of deep learning and neural networks, particularly convolutional neural networks (CNNs) for image recognition and recurrent neural networks (RNNs) for processing sequential data [61]. These technologies have revolutionized fields such as computer vision and natural language processing, enabling applications like real-time language translation and autonomous vehicle navigation [62].

AlphaGo's Victory Over World Go Champion

A landmark moment in AI history was when DeepMind's AlphaGo defeated Lee Sedol, a world champion Go player, in 2016. This achievement was significant due to the complexity of Go and demonstrated the advanced strategic capabilities of AI systems [9]. AlphaGo's success illustrated the potential of reinforcement learning and deep neural networks in solving complex problems.

Advancements in Natural Language Processing: GPT-3

The development of GPT-3 by OpenAI marked a major advancement in natural language processing. GPT-3, with its 175 billion parameters, demonstrated an unprecedented ability to generate human-like text, answering questions, writing essays, and even creating code [17]. This achievement showcased the potential of large-scale language models in understanding and generating human language.

Autonomous Vehicles and Robotics

The field of robotics has also seen significant AI achievements, particularly in the development of autonomous vehicles. Companies like Tesla and Waymo have made substantial progress in autonomous driving technology, contributing to advancements in sensor integration, real-time decision-making, and machine learning for navigation [63].

AI in Healthcare: Drug Discovery and Diagnosis

AI has made notable strides in healthcare, particularly in drug discovery and disease diagnosis. AI systems have been developed to analyze medical images with high accuracy, assist in diagnosing diseases like cancer, and even predict patient outcomes [64]. Additionally, AI has been used to accelerate the drug discovery process, identifying potential treatments for diseases more efficiently than traditional methods.

In conclusion, these notable achievements in AI demonstrate the technology's vast potential and its impact across various domains. From game-playing

AI to advancements in healthcare and autonomous vehicles, these achievements represent milestones in the journey of AI development and highlight the transformative power of this technology.

1.3.13   AI Development Across Different Countries

The development of artificial intelligence (AI) has become a key area of global competition, with various countries investing heavily in research and development to establish leadership in this transformative technology. This international race for AI supremacy is not only about technological advancement but also about economic power and geopolitical influence.

1.3.13.1   United States: A Pioneer in AI Research

The United States has been a pioneer in the field of AI, with its tech giants like Google, Apple, and Microsoft leading the way in AI research and development. The U.S. government has also played a significant role, with initiatives such as the American Artificial Intelligence Initiative, aimed at promoting and protecting national AI technology and innovation [65]. The combination of private sector innovation and public sector support has positioned the U.S. as a major player in the global AI landscape.

1.3.13.2   China: Ambitious Goals for AI Dominance

China has set ambitious goals to become the world leader in AI by 2030. The Chinese government's strategic plan for AI includes significant investments in AI research, education, and infrastructure. Chinese tech companies like Alibaba, Tencent, and Baidu are at the forefront of AI development in areas such as facial recognition technology and autonomous vehicles [22]. China's approach to AI is characterized by strong government support and a focus on practical applications.

1.3.13.3   European Union: Emphasizing Ethical AI

The European Union (EU) has focused on developing ethical guidelines for AI. The EU's approach emphasizes the responsible and human-centric development of AI, with policies and regulations designed to ensure privacy, transparency, and accountability in AI systems [11]. The EU's General Data Protection Regulation (GDPR) is an example of its commitment to regulating AI and data practices.

1.3.13.4   Other Key Players in AI Development

Other countries are also emerging as key players in AI development. Nations like South Korea, Japan, and Canada are investing in AI research and development, each with its unique strengths and focus areas. South Korea, for example, has made significant strides in robotics, while Canada is known for its expertise in deep learning and neural networks [62].

1.3.13.5   Global Collaboration and Competition

While there is intense competition in AI development among nations, there is also a growing recognition of the importance of global collaboration. International partnerships and collaborations in AI research are crucial for addressing global challenges and ensuring the responsible development of AI technologies [31].

In conclusion, AI development is a key area of global competition, with different countries pursuing their strategies and priorities. The United States and China are leading the race, but the EU and other nations are also making significant contributions. Balancing competition with collaboration will be essential for harnessing the full potential of AI for the benefit of all.

1.3.13.6   Notable AI Achievements Across Different Countries

The global landscape of artificial intelligence (AI) is marked by significant achievements from various countries, each contributing to the advancement of the field in unique ways. These achievements highlight the diverse approaches and strengths of different nations in the realm of AI.

United States: Pioneering AI Innovations

The United States has been at the forefront of many pioneering AI innovations. American tech giants like Google, Microsoft, and IBM have made groundbreaking advancements in areas such as deep learning, natural language processing, and cloud-based AI services. Google's development of TensorFlow, an opensource AI software library, and IBM's Watson, known for its performance in Jeopardy!, are notable examples [32, 66].

China: Rapid Advancements in AI Applications

China has made rapid advancements in practical AI applications. Chinese companies like Alibaba, Tencent, and Baidu have excelled in areas such as facial recognition technology, e-commerce, and autonomous vehicles. The country's focus on implementing AI in everyday life, from smart cities to education, showcases its commitment to becoming a global leader in AI [22].

European Union: Leading in Ethical AI Development

The European Union has been a leader in the ethical development of AI. With initiatives like GDPR and the establishment of various ethical guidelines for AI, the EU is at the forefront of ensuring responsible and human-centric AI development. The region's focus on balancing technological advancement with ethical considerations sets a precedent for the global AI community [11].

Canada: A Hub for AI Research

Canada is recognized as a hub for AI research, particularly in the field of deep learning. Canadian researchers and institutions, such as Geoffrey Hinton and the University of Toronto, have been instrumental in the development of neural network technology. The country's collaborative approach to AI research, involving academia, government, and industry, has fostered a thriving AI ecosystem [62].

Japan and South Korea: Innovations in Robotics and Automation

Japan and South Korea have been leaders in robotics and automation. Japanese companies like SoftBank and Sony have developed advanced humanoid robots, while South Korea has been a pioneer in industrial automation and robotics. These countries' focus on robotics reflects their long-term strategies to integrate AI into various aspects of society and industry [63].

In conclusion, the global achievements in AI reflect the diverse strengths and focuses of different countries. From pioneering innovations in the United States to rapid advancements in China, and from ethical AI development in the European Union to research excellence in Canada, these achievements collectively contribute to the global progress of AI technology.

1.3.14 Emerging Trends in AI Research

The field of artificial intelligence (AI) is continuously evolving, with new trends emerging that shape the future direction of research and development. These trends not only indicate the current focus areas in AI but also provide insights into how AI might evolve in the coming years.

1.3.14.1 Advancements in Machine Learning Algorithms

One of the key trends in AI research is the ongoing advancement of machine learning algorithms. Researchers are exploring beyond deep learning to develop new algorithms that are more efficient, require less data, and can make decisions with a higher degree of accuracy and explainability [67]. This includes work on meta-learning, where AI systems can learn how to learn, thereby improving their ability to adapt to new tasks.

1.3.14.2 AI and Ethics: Towards Responsible AI

Another significant trend is the increasing focus on ethical AI development. As AI systems become more prevalent, there is a growing need to ensure that they are developed and deployed responsibly. This includes addressing issues of bias, fairness, transparency, and accountability in AI systems [11]. Researchers are also exploring ways to incorporate ethical considerations directly into AI algorithms.

1.3.14.3 AI in Healthcare: Personalized Medicine and Drug Discovery

AI research in healthcare is rapidly advancing, with a focus on personalized medicine and drug discovery. AI systems are being developed to analyze patient data to provide personalized treatment recommendations and to assist in the discovery of new drugs and therapies [64]. This trend indicates the potential of AI to revolutionize healthcare by making it more effective and tailored to individual needs.

1.3.14.4 Quantum Computing and AI

The integration of quantum computing with AI is an emerging area of research. Quantum computing has the potential to significantly enhance the capabilities of AI systems, enabling them to solve complex problems much faster than traditional computers [68]. This convergence could lead to breakthroughs in various fields, including material science, finance, and cryptography.

1.3.14.5 AI and the Internet of Things (IoT)

The intersection of AI and the Internet of Things (IoT) is another growing trend. AI algorithms are being used to analyze data from IoT devices to make smarter decisions and automate tasks. This trend is leading to the development of smart cities, connected healthcare systems, and intelligent industrial processes [69].

In conclusion, the future directions of AI research are diverse and dynamic. From advancements in machine learning algorithms to the integration of AI with quantum computing and IoT, these trends indicate a future where AI will continue to have a profound impact on various aspects of society and industry.

1.3.14.6 Predictions and Speculations

The future of artificial intelligence (AI) is a subject of intense speculation and prediction, with experts and researchers offering various visions of what AI might bring in the coming years and decades. These predictions range from transformative impacts on society and industry to potential challenges and ethical dilemmas.

Transformation of Industries and Societal Impact

One common prediction is that AI will continue to transform industries, automating tasks and creating new efficiencies. This transformation is expected to impact sectors such as healthcare, finance, transportation, and manufacturing [49]. In healthcare, for instance, AI is anticipated to revolutionize diagnostics and personalized medicine, leading to better patient outcomes [64].

Advancements in General Artificial Intelligence

Another area of speculation is the development of general artificial intelligence (AGI) – AI systems that possess the ability to understand, learn, and apply knowledge in a way that is indistinguishable from human intelligence [70]. While the realization of AGI is still a matter of debate, its potential to redefine the capabilities of AI is a subject of significant interest.

Ethical and Societal Challenges

Predictions about AI also include potential ethical and societal challenges. Concerns about privacy, surveillance, job displacement, and decision-making biases in AI systems are prevalent in these discussions [43]. The need for robust ethical frameworks and regulations to guide the development and use of AI is a commonly echoed sentiment.

The Singularity and Superintelligence

The concept of the singularity – a point where AI surpasses human intelligence and becomes capable of self-improvement at an exponential rate – is a speculative scenario that has gained attention [52]. Associated with this is the notion of superintelligence, where AI systems become vastly more intelligent than humans, posing both opportunities and existential risks.

Long-Term Outlook and Human-AI Collaboration

In the long term, many experts speculate that the future of AI will be characterized by collaboration between humans and AI systems, rather than replacement. This collaboration is expected to enhance human capabilities and lead to the emergence of new forms of creativity and innovation [31].

In conclusion, while predictions and speculations about the future of AI vary widely, they all point to the profound impact AI is likely to have on our world. From transforming industries to raising new ethical questions, the future of AI is poised to be one of the most significant factors shaping the 21st century.

1.3.15 Case Study: DeepMind's AlphaGo

DeepMind's AlphaGo represents a landmark achievement in the field of artificial intelligence (AI). This case study delves into the development, achievements, and broader impact of AlphaGo, illustrating the capabilities and potential of AI in complex problem-solving.

1.3.15.1 Background and Development of AlphaGo

AlphaGo's journey began with the founding of DeepMind Technologies in London in 2010. DeepMind, co-founded by Demis Hassabis, Shane Legg, and Mustafa Suleyman, focused on creating a general-purpose AI [24]. The development of AlphaGo was part of this vision, aiming to tackle one of the most complex board games known to man: Go.

Go, an ancient Chinese board game, is known for its deep strategic elements and vast number of possible positions, making it a significant challenge for AI. Traditional AI methods that had succeeded in chess were inadequate for Go due to the game's complexity [71].

1.3.15.2 AlphaGo's Innovative Approach

AlphaGo's approach combined advanced machine learning techniques with Monte Carlo Tree Search (MCTS). It used deep neural networks, including convolutional neural networks, trained on thousands of human amateur and professional Go games to learn patterns and strategies [9]. This training allowed AlphaGo to predict moves and evaluate board positions effectively.

1.3.15.3 Historic Match Against Lee Sedol

The most notable moment in AlphaGo's history was its match against Lee Sedol, one of the world's top Go players, in March 2016. The match was a five-game challenge held in Seoul, South Korea. AlphaGo's victory in four out of five games was a watershed moment for AI, demonstrating that machine learning could outperform human intelligence in complex tasks [24].

1.3.15.4 Impact and Implications of AlphaGo's Victory

AlphaGo's victory had far-reaching implications. It was not just a triumph in the realm of board games but also a demonstration of the potential of AI in solving complex, real-world problems. The techniques developed for AlphaGo have potential applications in areas such as drug discovery, climate modeling, and complex system optimization [72].

1.3.15.5 AlphaGo's Legacy and Continued Evolution

The legacy of AlphaGo extends beyond its match with Lee Sedol. DeepMind continued to develop the program, leading to the creation of AlphaGo Zero, which learned to play Go without human data, purely through self-play reinforcement learning [12]. This advancement marked another significant step in AI research, showing that AI could achieve superhuman performance without human input.

In conclusion, DeepMind's AlphaGo is a seminal case in the history of AI, symbolizing the field's rapid advancement and potential. From its innovative approach to learning and problem-solving to its historic achievements and ongoing evolution, AlphaGo exemplifies the transformative power of artificial intelligence.

1.3.16 Governmental Policies on AI

The rapid advancement of artificial intelligence (AI) technologies has prompted governments around the world to develop policies and regulations to govern their development and use. These policies are crucial for ensuring that AI technologies are developed and deployed in a safe, ethical, and beneficial manner.

1.3.16.1 United States: A Focus on Innovation and Research

In the United States, AI policy has largely focused on promoting innovation and research. The American AI Initiative, launched in 2019, aims to accelerate AI development and regulation by increasing investment in AI research, promoting open data and standards, and training the future AI workforce [65]. This initiative reflects the U.S. government's approach to maintain leadership in AI technology while addressing its societal impacts.

1.3.16.2 European Union: Emphasizing Ethical AI

The European Union (EU) has taken a proactive approach in regulating AI, with a strong emphasis on ethics. The EU's guidelines for trustworthy AI focus on ensuring that AI systems are lawful, ethical, and robust. The General Data Protection Regulation (GDPR), implemented in 2018, is a key component of this approach, providing regulations on data privacy and protection, which directly impact AI development [11].

1.3.16.3 China: State-Led AI Development

China's approach to AI policy is characterized by significant state involvement and investment. The Chinese government's "New Generation Artificial Intelligence Development Plan" aims to make China a global leader in AI by 2030. This plan includes substantial funding for AI research, incentives for AI companies, and integration of AI into various sectors of the economy [22].

1.3.16.4 Global Collaboration and Standards

There is also a growing recognition of the need for global collaboration and standards in AI policy. International organizations like the OECD and the G7 have begun to develop guidelines and principles for the responsible stewardship of trustworthy AI. These efforts aim to harmonize AI policies across different countries and promote international cooperation in the development of AI [31].

1.3.16.5 Challenges in AI Regulation

Developing effective AI policies poses several challenges. These include balancing innovation with ethical considerations, addressing the global and rapidly evolving nature of AI technology, and dealing with issues such as data privacy, security, and the potential for bias in AI systems. Governments must navigate these challenges to create policies that foster the responsible development and use of AI.

In conclusion, governmental policies on AI are crucial for guiding the development and deployment of AI technologies. Different countries have adopted various approaches, reflecting their unique priorities and values. As AI continues to advance, these policies will play a key role in shaping the future of AI and its impact on society.

1.3.16.6 Global Regulatory Landscape

The global regulatory landscape for artificial intelligence (AI) is diverse and evolving, with various countries and international organizations developing frameworks and guidelines to govern the ethical development and deployment of AI technologies.

United Nations and International Guidelines

The United Nations (UN) and other international bodies have been instrumental in shaping the global discourse on AI regulation. The UN's initiatives, such as the International Telecommunication Union's AI for Good Global Summit, aim to establish a common understanding of AI ethics and standards [73]. Additionally, the OECD's Principles on Artificial Intelligence set out guidelines for responsible stewardship of trustworthy AI, which have been adopted by many countries [74].

European Union's Leadership in AI Regulation

The European Union (EU) has been a leader in the global regulatory landscape for AI. The EU's approach is characterized by its focus on ethical and human-centric AI. The General Data Protection Regulation (GDPR) is a cornerstone of this approach, providing a framework for data protection and privacy that impacts AI development [75]. The EU has also proposed regulations specifically for AI, focusing on high-risk applications and fundamental rights [76].

United States' Approach to AI Regulation

In the United States, AI regulation is primarily focused on fostering innovation and competitiveness. The U.S. government has issued guidelines for federal agencies to promote the development of AI technologies while ensuring public engagement, limiting regulatory overreach, and promoting trustworthy AI [77].

This approach reflects the U.S.'s emphasis on maintaining technological leadership.

China's State-Led AI Strategy

China's approach to AI regulation is part of its broader state-led strategy to become a global leader in AI by 2030. The Chinese government's New Generation Artificial Intelligence Development Plan outlines a comprehensive strategy for AI development, including ethical guidelines, standards, and policies to promote AI innovation while ensuring security and privacy [78].

Diverse Approaches in Other Countries

Other countries, such as Canada, Japan, South Korea, and Singapore, have also developed their AI policies and frameworks. These range from national strategies to promote AI research and development to guidelines on ethical AI use. The diversity of approaches reflects the varying priorities and values of different countries in the global AI landscape [79].

In conclusion, the global regulatory landscape for AI is characterized by a patchwork of approaches, reflecting the diverse priorities and values of different countries and international organizations. As AI continues to advance, this landscape will likely evolve, with ongoing discussions and collaborations shaping the future of AI regulation.

1.4 Dramatic Moments and Key Discussions in AI's Evolution

The development of artificial intelligence (AI) has been marked not only by technological breakthroughs but also by dramatic moments, secret gatherings, and pivotal discussions among tech giants. This section highlights these exciting aspects as detailed in the New York Times article.

1.4.1 Secret Gatherings and Elite Meetings

The AI journey is dotted with secretive and exclusive meetings where tech leaders and AI pioneers exchange groundbreaking ideas. One such instance was Elon Musk's 44th birthday party in July 2015, held at a California wine country resort. This gathering, which included guests like Larry Page, was a blend of personal celebration and high-level tech discussions [80].

1.4.2 Elon Musk and Larry Page: A Clash of Visions

A particularly captivating aspect of these gatherings is the discussions between figures like Elon Musk and Larry Page. Musk, known for his cautionary stance on AI, often finds himself in debate with Page, who holds a more optimistic view. Their discussions, sometimes stretching into the night, revolve around whether AI will ultimately benefit humanity or pose a significant threat [80].

1.4.3 The Backstage of AI Development

The article also sheds light on the intense research and unexpected breakthroughs in AI. For instance, the development of systems like AlphaGo involved moments of triumph that surprised even its creators. These stories highlight the dynamic nature of AI research, where each development is a step into uncharted territory [80].

1.4.4 The Human Element in AI Evolution

Amidst the technological advancements, the human element remains pivotal. The passion, curiosity, and sometimes rivalry among AI researchers are key drivers of the field. The article captures this human element, showcasing how personal visions, ethical considerations, and collaborative efforts shape the trajectory of AI development [80].

In conclusion, the world of AI is as much about the people behind the technology as it is about the technology itself. From secret dinners to visionary debates and the intense pursuit of knowledge, these highlights provide a vivid picture of the AI frontier, full of excitement, challenges, and the promise of a transformative future.

1.5 Conclusion

1.5.1 Summary of Key Points

This chapter has explored various facets of artificial intelligence (AI), from its historical context to the latest advancements and global impact. Key points include:

- **Historical Context of AI:** The evolution of AI from its early days to the present, highlighting significant milestones.

- **Notable AI Systems and Technologies:** An overview of groundbreaking AI systems like DeepMind's AlphaGo and emerging technologies in the field.

- **Global AI Race:** The competitive dynamics of AI development across different countries, emphasizing the varied approaches and strategies.

- **Ethical Considerations and Societal Impact:** The ethical dilemmas and societal implications of AI, including debates on privacy, bias, and automation's impact on employment.

- **Regulatory and Policy Considerations:** A look at how different governments and international bodies are shaping AI development through policies and regulations.

- **Exciting Anecdotes from the AI World:** Highlights from the New York Times article, capturing the human stories and dramatic moments in the AI journey.

1.5.2 Future Research Directions

Looking ahead, the field of AI presents numerous avenues for future research:

- **Advancing AI Technologies:** Continued research in deep learning, neural networks, and emerging areas like quantum AI and meta-learning.

- **Ethical AI Development:** Further exploration of how to integrate ethical considerations into AI algorithms and systems.

- **AI in Healthcare:** Expanding AI's role in personalized medicine, drug discovery, and healthcare management.

- **Human-AI Collaboration:** Investigating how humans and AI can work together effectively, enhancing human capabilities without replacing them.

- **Global AI Governance:** Developing international standards and collaborative frameworks for responsible AI development and use.

In conclusion, artificial intelligence stands at the forefront of technological advancement, offering both immense potential and significant challenges. As the field continues to evolve, it will be crucial to balance innovation with ethical considerations, ensuring that AI benefits society as a whole. The future of AI is not just about technological prowess but also about the wisdom with which we guide its development and integration into our world.

Chapter 2

From Turing to Transformers: A Comprehensive Review and Tutorial on the Evolution and Applications of Generative Transformer Models

---

2.1 Introduction

2.1.1 Background and Significance of Generative Models in AI

Generative models serve as an essential building block in the realm of artificial intelligence (AI). At their core, these models are designed to generate new data samples that are similar to the input data they have been trained on. This capability has profound implications, enabling machines to create, imagine, and replicate complex patterns observed in the real world.

The inception of generative models can be traced back to the early days of AI, where the foundational work of Alan Turing laid the groundwork for the evolution of generative models and the broader field of AI. Following Turing's pioneering contributions, the field witnessed the emergence of simple algorithms designed to mimic and reproduce sequential data. An exemplar of this era is the Hidden Markov Models (HMM) proposed by Leonard Baum in a series of seminal papers published in the late 1960s [81, 82, 83]. These models were groundbreaking for their time, providing a probabilistic framework to understand and predict sequences. The most notable application of HMMs was in the realm of speech recognition [84], where they became a foundational component, enabling systems to decode and understand human speech with increasing accuracy.

The introduction of Recurrent Neural Networks (RNNs) in 1982 by John Hopfield [85] and Long Short-Term Memory (LSTM) networks in 1997 by Hochreiter and Schmidhuber [86] marked significant advancements in the field. RNNs brought the ability to remember previous inputs in handling sequential data, while LSTMs addressed the challenges of long-term dependencies, making them pivotal for tasks like time series prediction, speech recognition, and natural language processing. Together, they set foundational standards for modern generative AI models handling sequences.

However, with the advent of deep learning and the proliferation of neural networks, the potential and capabilities of generative models have expanded exponentially. Neural-based generative models, such as Variational Autoencoders (VAEs) [87, 88] introduced in 2013 and Generative Adversarial Networks (GANs) [89, 90] introduced in the following year, have showcased the ability to generate high-fidelity new data samples based on training data, ranging from images to text and even music.

The significance of generative models in AI is multifaceted. Firstly, they play a pivotal role in unsupervised learning, where labeled data is scarce or unavailable. By learning the underlying distribution of the data, generative models can produce new samples, aiding in tasks like data augmentation [91, 92], anomaly detection [93], and image denoising [94, 95]. Secondly, the creative potential of these models has been harnessed in various domains, from image [96, 97, 98, 99], video and music generation to drug discovery [100, 101] and virtual reality [102, 103, 104]. The ability of machines to generate novel and coherent content has opened up avenues previously deemed exclusive to human creativity.

Furthermore, generative models serve as powerful tools for understanding and interpreting complex data distributions. They provide insights into the structure and relationships within the data, enabling researchers and practitioners to uncover hidden patterns, correlations, and features [105]. This interpretative power is especially valuable in domains like biology [106], finance [107], and climate science [108], where understanding data intricacies can lead to ground-breaking discoveries.

Generative models stand as a testament to the advancements and possibilities within AI. Their ability to create, interpret, and innovate has not only broadened the horizons of machine learning but has also reshaped our understanding of intelligence and creativity.

2.1.2 The Rise of Transformer Architectures

While Variational Autoencoders (VAEs) and Generative Adversarial Networks (GANs) have significantly advanced the field of generative AI, another monumental shift in the deep learning landscape emerged with the introduction of the transformer architecture. Presented in the seminal paper "Attention is All You Need" by a team of Google researchers led by Vaswani in 2017 [109], transformers have redefined the benchmarks in a multitude of tasks, particularly in natural language processing (NLP).

The transformer's innovation lies in its self-attention mechanism, which allows it to weigh the significance of different parts of an input sequence, be it words in a sentence or pixels in an image. This mechanism enables the model to capture long-range dependencies and intricate relationships in the data, overcoming the limitations of previous architectures like Recurrent Neural Networks (RNNs) and Long Short-Term Memory (LSTM) networks. RNNs and LSTMs, while effective in handling sequential data, often struggled with long sequences due to issues like vanishing and exploding gradients [110]. Transformers, with their parallel processing capabilities and attention mechanisms, alleviated these challenges.

The success of the transformer architecture was not immediate but became evident with the introduction of large language models like BERT (Bidirectional Encoder Representations from Transformers) and GPT (Generative Pre-trained Transformer). BERT, developed by researchers at Google, demonstrated the power of transformers in understanding the context of words in a sentence by considering both left and right contexts in all layers [111]. This bidirectional approach led to state-of-the-art results in several NLP tasks, from question answering to sentiment analysis [112]. On the other hand, OpenAI's GPT showcased the generative capabilities of transformers [113], producing human-like text and achieving remarkable performance in tasks like machine translation [114] and text summarization [115] without task-specific training data.

The transformer's versatility extends beyond NLP. Vision Transformer (ViT) [116], an adaptation of the architecture for image classification tasks, has shown that transformers can rival, if not surpass, the performance of traditional convolutional neural networks (CNNs) in computer vision tasks [117, 118]. This cross-domain applicability underscores the transformer's potential and its foundational role in modern AI.

Another driving factor behind the rise of transformers is the ever-growing computational power and the availability of large-scale datasets. Training transformer models, especially large ones, requires significant computational resources. The feasibility of training such models has been made possible due to advancements in GPU and TPU technologies [119], coupled with the availability of vast amounts of data to train on. The combination of innovative architecture and computational prowess has led to the development of models with billions or even trillions of parameters, pushing the boundaries of what machines can generate to new heights.

Generative AI models have undergone significant transformations since their inception, with each milestone contributing to the capabilities we see today. From the foundational Turing machines to the latest GPT-4 and LLaMA models, the journey of generative AI has been marked by groundbreaking advancements. A detailed timeline capturing these key milestones is presented to offer a comprehensive overview of the field's evolution (Figure 2.1).

2.1.3 Purpose and Structure of the Chapter

The fast growth in artificial intelligence, especially with recent technologies like generative models and transformers, highlights the need for a comprehensive study that spans both their historical development and current applications. The primary objective of this paper is to provide readers with a holistic understanding of the evolution, significance, architecture, and capabilities of generative transformers, contextualized within the broader landscape of AI.

Our motivation for this paper is informed by the existing body of work on transformer-based models and generative AI. While there are several comprehensive reviews, each focuses on specific aspects of the topic. For example, Gozalo-Brizuela and Garrido-Merchan [120] concentrate on the taxonomy and industrial implications of large generative models, providing a compilation of popular generative models organized into various categories such as text-to-text, text-to-image, and text-to-audio. Lin et al. [121] present an exhaustive review of various transformer variants, their architectural modifications, and applications. Additionally, there are survey papers that focus on the use of transformers for specific tasks such as natural language processing [122, 123], computer vision [124, 125, 126, 127], time series analysis and forecasting [128, 129], among others. These existing reviews are invaluable, but our paper aims to provide a more comprehensive overview that bridges these specialized areas.

While these papers offer valuable insights, there is a gap in the literature for a resource that combines a historical review, a hands-on tutorial, and a forward-looking perspective on generative transformer models. Our paper aims to fill this void, serving as a comprehensive guide for newcomers and seasoned researchers alike. The historical review section helps readers understand how generative AI has developed and progressed in the wider context of AI. Meanwhile, our prac-

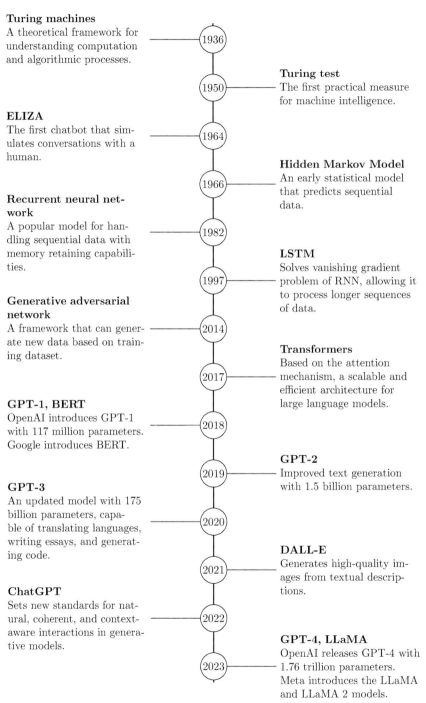

Turing machines
A theoretical framework for understanding computation and algorithmic processes.

1936

Turing test
The first practical measure for machine intelligence.

1950

ELIZA
The first chatbot that simulates conversations with a human.

1964

Hidden Markov Model
An early statistical model that predicts sequential data.

1966

Recurrent neural network
A popular model for handling sequential data with memory retaining capabilities.

1982

LSTM
Solves vanishing gradient problem of RNN, allowing it to process longer sequences of data.

1997

Generative adversarial network
A framework that can generate new data based on training dataset.

2014

Transformers
Based on the attention mechanism, a scalable and efficient architecture for large language models.

2017

GPT-1, BERT
OpenAI introduces GPT-1 with 117 million parameters. Google introduces BERT.

2018

GPT-2
Improved text generation with 1.5 billion parameters.

2019

GPT-3
An updated model with 175 billion parameters, capable of translating languages, writing essays, and generating code.

2020

DALL-E
Generates high-quality images from textual descriptions.

2021

ChatGPT
Sets new standards for natural, coherent, and context-aware interactions in generative models.

2022

GPT-4, LLaMA
OpenAI releases GPT-4 with 1.76 trillion parameters. Meta introduces the LLaMA and LLaMA 2 models.

2023

Figure 2.1: A timeline illustrating key milestones in the development of generative AI, from Turing Machines to GPT-4.

tical tutorial guides readers through the foundational concepts and practical implementations, equipping them to build their own generative transformer models. We offer a unique blend of theoretical understanding and practical know-how, setting our work apart from existing reviews. Additionally, we strive to provide a unique balance between explaining the historical evolution, technical aspects, and applications of transformers. This makes our paper a go-to source for researchers and professionals seeking a wholesome understanding and knowledge of transformers.

The structure of the paper, which is designed to guide the reader through a logical progression, is as follows:

- **Historical Evolution:** We embark on a journey tracing the roots of computational theory, starting with the foundational concepts introduced by Alan Turing. This section provides a backdrop, setting the stage for the emergence of neural networks, the challenges they faced, and the eventual rise of transformer architectures.

- **Tutorial on Generative Transformers:** Transitioning from theory to practice, this section offers a practical approach to understanding the intricacies of generative transformers. Readers will gain insights into the architecture, training methodologies, and best practices, supplemented with code snippets and practical examples.

- **Applications and Challenges:** Building upon the foundational knowledge, we delve into the myriad applications of generative transformers, highlighting their impact across various domains. Concurrently, we address the challenges and ethical considerations associated with their use, fostering a balanced perspective.

- **Conclusion and Future Directions:** The paper concludes with a reflection on the current state of generative transformers, their potential trajectory, and the exciting possibilities they hold for the future of AI.

In essence, this paper endeavors to be more than just a review or a tutorial, it aspires to be a comprehensive guide, weaving together history, theory, practice, and prospects, providing readers with a panoramic view of the world of generative transformers.

2.2 Historical Evolution

The development of computational theory and artificial intelligence has been shaped by pioneering figures, innovative ideas, and transformative discoveries. Central to this narrative is Alan Turing, whose unparalleled contributions laid the foundations for modern computation and the subsequent emergence of AI.

This section delves deeper into Turing's groundbreaking work, and the lasting legacy that continues to shape the digital age.

2.2.1 Turing Machines and the Foundations of Computation

One of Turing's major contributions was the idea of the Turing machine proposed in his 1936 paper titled "On Computable Numbers, with an Application to the Entscheidungsproblem." [130] This abstract machine was a simple but powerful theoretical construct that was designed to perform computations by manipulating symbols on an infinite tape based on a set of rules. The infinite tape is divided into discrete cells, each cell can contain a symbol from a finite alphabet, and the machine itself has a "head" that can read and write symbols on the tape and move left or right. The machine's behavior is dictated by a set of transition rules, which determine its actions based on the current state and the symbol being read. In essence, the Turing machine is a rule-based system that manipulates symbols on a tape, embodying the fundamental operations of reading, writing, and transitioning between states.

While the concept might seem rudimentary, the implications of the Turing machine are profound. Turing demonstrated that this simple device, with its set of rules and operations, could compute any function that is computable, given enough time and tape. This assertion, known as the Church-Turing thesis [131] (independently proposed by Alonzo Church in his paper titled "An Unsolvable Problem of Elementary Number Theory" also published in 1936 [132]), posits that any function computable by an algorithm can be computed by a Turing machine. This thesis, though not proven, has stood the test of time, with no evidence to the contrary. It serves as a foundational pillar in computer science, defining the boundaries of what is computable.

World War II saw Turing's theoretical concept manifest in tangible, real-world applications. Stationed at Bletchley Park, Britain's cryptographic hub, Turing played a key role in deciphering the Enigma code used by the German military. Turing helped develop a machine called the Bombe, which expedited the decryption process of Enigma-encrypted messages [133]. This secret work was crucial for the Allies' success and showed how computer science could have a major impact on real-world events.

After World War II, Turing turned his attention to the development of electronic computers. He was instrumental in the design of the Automatic Computing Engine (ACE) [134], one of the earliest computer models capable of storing programs. This showed Turing's forward-thinking approach to the digital age. Beyond computing, he also delved into the nature of intelligence and how it could be replicated in machines.

The Turing machine's significance transcended its immediate mathematical implications. The true brilliance of Turing's insight, however, lies in the concept of universal computation. Turing's subsequent proposition of a Universal

Turing Machine (UTM)—a machine capable of simulating any other Turing machine given the right input and rules—was a revolutionary idea [130]. Given a description of a Turing machine and its input encoded on the tape, the UTM could replicate the behavior of that machine. This meta-level of computation was groundbreaking. It suggested that a single, general-purpose machine could be designed to perform any computational task, eliminating the need for task-specific machines. The UTM was a harbinger of modern computers, devices that can be reprogrammed to execute a wide array of tasks.

The implications of universal computation extend beyond mere hardware. It challenges our understanding of intelligence and consciousness. If the human brain, with its intricate neural networks and synaptic connections, operates on computational principles, then could it be simulated by a Turing machine? This question, which blurs the lines between philosophy, neuroscience, and computer science, remains one of the most intriguing and debated topics in the field of artificial intelligence.

2.2.1.1 *Turing's Impact on Artificial Intelligence and Machine Learning*

Alan Turing's influence on the fields of artificial intelligence (AI) and machine learning (ML) is both profound and pervasive. While Turing is often lauded for his foundational contributions to computational theory, his vision and insights into the realm of machine intelligence have played a pivotal role in shaping the trajectory of AI and ML.

His 1950 paper, "Computing Machinery and Intelligence," [135] introduced the famous Turing Test as a practical measure of machine intelligence. Alan Turing introduced the Turing Test within the context of an "Imitation Game," involving a man, a woman, and a judge as players. They communicate electronically from separate rooms, and the goal of the judge is to identify who is the woman. The man aims to deceive the judge into thinking he is the woman, while the woman assists the judge. Turing then adapts this game into his famous test by replacing the man with a machine, aiming to deceive the questioner in the same way. Although the original game focused on gender identification, this aspect is often overlooked in later discussions of the Turing Test.

In this work, Turing posed the provocative question: "Can machines think?" Rather than delving into the philosophical intricacies of defining "thinking," Turing proposed a pragmatic criterion for machine intelligence: if a machine could engage in a conversation with a human, indistinguishably from another human, it would be deemed intelligent. This criterion, while straightforward, sparked widespread debate and research, laying the foundation for the field of artificial intelligence.

The Turing Test, in many ways, encapsulated the essence of AI—the quest to create machines that can mimic, replicate, or even surpass human cognitive abilities. It set a benchmark, a gold standard for machine intelligence, challenging researchers and scientists to build systems that could "think" and "reason" like humans. While the test itself has been critiqued and refined over the years, its underlying philosophy remains central to AI: the aspiration to understand and emulate human intelligence.

Beyond the Turing Test, Turing's insights into neural networks and the potential of machine learning were visionary. In a lesser-known report written in 1948, titled "Intelligent Machinery," [136] Turing delved into the idea of machines learning from experience. He envisioned a scenario where machines could be trained, much like a human child, through a process of education. Turing postulated the use of what he termed "B-type unorganized machines," which bear a striking resemblance to modern neural networks. These machines, as Turing described, would be trained, rather than explicitly programmed, to perform tasks. Although in its infancy at the time, this idea signaled the rise of machine learning, where algorithms learn from data rather than being explicitly programmed.

Turing's exploration of morphogenesis, the biological process that causes organisms to develop their shape, further showcased his interdisciplinary genius [137]. In his work on reaction-diffusion systems, Turing demonstrated how simple mathematical models could give rise to complex patterns observed in nature. This work, while primarily biological in its focus, has profound implications for AI and ML. It underscores the potential of simple algorithms to generate complex, emergent behavior, a principle central to neural networks and deep learning.

Alan Turing's impact on artificial intelligence and machine learning is immeasurable. His vision of machine intelligence, his pioneering insights into neural networks, and his interdisciplinary approach to problem-solving have left an indelible mark on the field. As we navigate the intricate landscape of modern AI, with its deep neural networks, generative models, and transformers, it is imperative to recognize and honor Turing's legacy. His work serves as a beacon, illuminating the path forward, reminding us of the possibilities, challenges, and the profound potential of machines that can "think."

2.2.1.2 From Turing's Foundations to Generative Transformers

The journey from Alan Turing's foundational concepts to the sophisticated realm of generative transformers is a testament to the evolution of computational theory and its application in artificial intelligence. While at first glance Turing's work and generative transformers might seem worlds apart, a closer examination reveals a direct lineage and influence.

Alan Turing's conceptualization of the Turing machine provided the bedrock for understanding computation. His idea of a machine that could simulate any algorithm, given the right set of instructions, laid the groundwork for the concept of universal computation. This idea, that a single machine could be reprogrammed to perform a myriad of tasks, is the precursor to the modern notion of general-purpose computing systems.

Fast forward to the advent of neural networks, which Turing had touched upon in his lesser-known works. These networks, inspired by the human brain's interconnected neurons, were designed to learn from data. The foundational idea was that, rather than being explicitly programmed to perform a task, these networks would "learn" by adjusting their internal parameters based on the data they were exposed to. Turing's vision of machines learning from experience resonates deeply with the principles of neural networks.

Generative transformers, a cutting-edge development in the AI landscape, are an extension of these neural networks. Transformers, with their self-attention mechanisms, are designed to weigh the significance of different parts of an input sequence, capturing intricate relationships within the data. The "generative" aspect of these models allows them to produce new, previously unseen data samples based on their training.

Drawing a direct link, Turing's Universal Turing Machine can be seen as an early, abstract representation of what generative transformers aim to achieve in a more specialized domain. Just as the Universal Turing Machine could simulate any other Turing machine, given the right input and set of rules, generative transformers aim to generate any plausible data sample, given the right training and context. The universality of Turing's machine finds its parallel in the versatility of generative transformers.

Furthermore, Turing's exploration into machine learning, the idea of machines learning from data rather than explicit programming, is the very essence of generative transformers. These models are trained on vast datasets, learning patterns, structures, and nuances, which they then use to generate new content. The bridge between Turing's early insights into machine learning and the capabilities of generative transformers is a direct one, showcasing the evolution of a concept from its theoretical inception to its practical application.

While Alan Turing might not have directly worked on generative transformers, his foundational concepts, vision of machine learning, and the principles he laid down have directly influenced and shaped their development. The journey from Turing machines to generative transformers is a testament to the enduring legacy of Turing's genius and the continual evolution of artificial intelligence.

2.2.2   Early Neural Networks and Language Models

The realm of artificial intelligence has witnessed a plethora of innovations and advancements, with neural networks standing at the forefront of this revolution. These computational models, inspired by the intricate web of neurons in the human brain, have paved the way for sophisticated language models that can understand, generate, and manipulate human language with unprecedented accuracy.

2.2.2.1 Introduction to Neural Networks

Neural networks [138, 139], at their core, are a set of algorithms designed to recognize patterns. They interpret sensory data through a kind of machine perception, labeling, and clustering of raw input. These algorithms loosely mirror the way a human brain operates, thus the nomenclature "neural networks."

A basic neural network consists of layers of interconnected nodes or "neurons." Each connection between neurons has an associated weight, which is adjusted during training. The fundamental equation governing the output y of a neuron is given by:

$$y = f\left(\sum_i w_i x_i + b\right) \tag{2.1}$$

where x_i are the input values, w_i are the weights, b is a bias term, and f is an activation function.

The activation function introduces non-linearity into the model, allowing it to learn from error and make adjustments, which is essential for learning complex patterns. One of the commonly used activation functions is the sigmoid function, defined as:

$$f(z) = \frac{1}{1 + e^{-z}} \tag{2.2}$$

Neural networks typically consist of an input layer, one or more hidden layers, and an output layer. The depth and complexity of a network, often referred to as its "architecture," determine its capacity to learn from data.

2.2.2.2 Evolution of Recurrent Neural Networks (RNNs)

While traditional neural networks have proven effective for a wide range of tasks, they possess inherent limitations when dealing with sequential data. This is where Recurrent Neural Networks (RNNs) come into play. RNNs are designed to recognize patterns in sequences of data, such as time series or natural language.

The fundamental difference between RNNs and traditional neural networks lies in the former's ability to retain memory of previous inputs in its internal state. This is achieved by introducing loops in the network, allowing information to persist.

The output of a RNN at time t, denoted h_t, is computed as:

$$h_t = f\left(W_{hh} h_{t-1} + W_{xh} x_t + b\right) \tag{2.3}$$

where W_{hh} and W_{xh} are weight matrices, x_t is the input at time t, and h_{t-1} is the output from the previous timestep.

While RNNs are powerful, they suffer from challenges like the vanishing and exploding gradient problems, especially when dealing with long sequences [110]. This makes them less effective in capturing long-term dependencies in the data.

2.2.2.3 Long Short-Term Memory (LSTM) Networks

To address the vanishing gradient problem of RNNs, Long Short-Term Memory (LSTM) networks were introduced. LSTMs, a special kind of RNN, are designed to remember information for extended periods [140].

The core idea behind LSTMs is the cell state, a horizontal line running through the entire chain of repeating modules in the LSTM. The cell state can carry information from earlier time steps to later ones, mitigating the memory issues faced by traditional RNNs.

LSTMs introduce three gates:

1. **Forget Gate**: It decides what information from the cell state should be thrown away or kept. Mathematically, the forget gate f_t is given by:

$$f_t = \sigma(W_f \cdot [h_{t-1}, x_t] + b_f) \tag{2.4}$$

2. **Input Gate**: It updates the cell state with new information. The input gate i_t and the candidate values $\tilde{C}_t$ are computed as:

$$i_t = \sigma(W_i \cdot [h_{t-1}, x_t] + b_i) \tag{2.5}$$

$$\tilde{C}_t = \tanh(W_C \cdot [h_{t-1}, x_t] + b_C) \tag{2.6}$$

3. **Output Gate**: It determines the output based on the cell state and the input. The output h_t is given by:

$$h_t = o_t \times \tanh(C_t) \tag{2.7}$$

where o_t is the output gate, defined as:

$$o_t = \sigma(W_o \cdot [h_{t-1}, x_t] + b_o) \tag{2.8}$$

LSTMs, with their ability to capture long-term dependencies and mitigate the challenges faced by traditional RNNs, have paved the way for advancements in sequence modeling, particularly in the domain of natural language processing.

2.2.3 The Advent of Transformers

In the ever-evolving landscape of artificial intelligence and machine learning, the transformer architecture stands out as a significant leap forward, especially in the domain of natural language processing. Introduced in the seminal paper "Attention Is All You Need" by Vaswani et al. [109], transformers have revolutionized the way we approach sequence-to-sequence tasks. This section aims to demystify the transformer architecture, breaking it down into its core components and principles.

2.2.3.1 Introduction to the Transformer Architecture

At a high level, the transformer is a type of neural network architecture designed to handle sequential data, making it particularly well-suited for tasks like language translation, text generation, and more. Unlike its predecessors, such as RNNs and LSTMs, which process data in order, transformers leverage a mechanism called "attention" to draw global dependencies between input and output.

The Attention Mechanism:

The heart of the transformer architecture is the attention mechanism. In essence, attention allows the model to focus on different parts of the input sequence when producing an output sequence, much like how humans pay attention to specific words when understanding a sentence.

Mathematically, the attention score for a given query q and key k is computed as:

$$\text{Attention}(q,k) = \frac{\exp(\text{score}(q,k))}{\sum_{k'} \exp(\text{score}(q,k'))} \tag{2.9}$$

where score is a function that calculates the relevance of the key k to the query q. The output of the attention mechanism is a weighted sum of values, where the weights are the attention scores.

The Transformer Architecture:

The transformer model consists of an encoder and a decoder. Each of these is composed of multiple layers of attention and feed-forward neural networks.

The encoder takes in a sequence of embeddings (representations of input tokens) and processes them through its layers. The decoder then generates the output sequence, leveraging both its internal layers and the encoder's output.

One of the distinguishing features of transformers is the use of "multi-head attention," which allows the model to focus on different parts of the input simultaneously, capturing various aspects of the information.

Why Transformers?

■ **Parallelization**: Unlike RNNs, which process sequences step-by-step, transformers can process all tokens in parallel, leading to faster training times.

■ **Long-range Dependencies**: The attention mechanism enables transformers to capture relationships between tokens, regardless of their distance in the sequence.

■ **Scalability**: Transformers are highly scalable, making them suitable for large datasets and complex tasks.

The transformer architecture, with its innovative attention mechanism and parallel processing capabilities, has set new benchmarks in the field of machine learning. Its ability to capture intricate patterns and relationships in sequential data has paved the way for state-of-the-art models in natural language processing, making tasks like real-time translation, text summarization, and question-answering more accurate and efficient.

2.2.4   Attention Mechanism: The Heart of Transformers

The attention mechanism, a pivotal innovation in the realm of deep learning, has transformed the way we approach sequence-to-sequence tasks in natural language processing. Serving as the cornerstone of the transformer architecture, attention allows models to dynamically focus on different parts of the input data, capturing intricate relationships and dependencies. This section aims to elucidate the principles and mathematics behind the attention mechanism, shedding light on its significance in the transformer architecture.

2.2.4.1   Conceptual Overview of Attention

In traditional sequence-to-sequence models, such as RNNs and LSTMs, information from the entire input sequence is compressed into a fixed-size context vector, which is then used to generate the output sequence. This approach, while effective for short sequences, struggles with longer sequences as the context vector becomes a bottleneck, unable to capture all the nuances of the input data.

The attention mechanism addresses this challenge by allowing the model to "attend" to different parts of the input sequence dynamically, based on the current context. Instead of relying on a single context vector, the model computes a weighted sum of all input vectors, where the weights represent the "attention scores."

2.2.4.2   Mathematics of Attention

The core of the attention mechanism is the computation of attention scores. Given a query q and a set of key-value pairs (k, v), the attention score for a specific key k is computed as:

$$\text{score}(q, k) = q^T k \tag{2.10}$$

The attention weights, which determine how much focus should be given to each key-value pair, are computed using a softmax function:

$$\text{Attention}(q, k) = \frac{\exp(\text{score}(q, k))}{\sum_{k'} \exp(\text{score}(q, k'))} \tag{2.11}$$

The output of the attention mechanism is a weighted sum of the values:

$$\text{output} = \sum_i \text{Attention}(q, k_i) v_i \qquad (2.12)$$

As depicted in Figure 2.2, the attention mechanism computes scores based on the query and keys, derives attention weights, and produces an output based on a weighted sum of values.

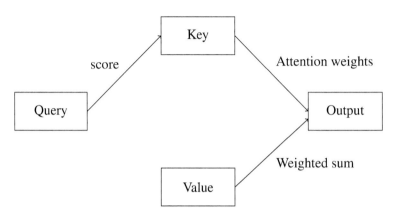

Figure 2.2: Schematic representation of the attention mechanism.

2.2.4.3 Significance in Transformers

In the transformer architecture, attention is not just a supplementary feature; it's the core component. Transformers employ a variant called "multi-head attention," which runs multiple attention mechanisms in parallel, capturing different types of relationships in the data.

The attention mechanism's ability to focus on different parts of the input sequence, irrespective of their position, empowers transformers to handle long-range dependencies, making them particularly effective for tasks like language translation, text summarization, and more.

Furthermore, the self-attention mechanism, a special case where the query, key, and value are all derived from the same input, enables transformers to weigh the significance of different parts of the input relative to a specific position. This is crucial for understanding context and semantics in natural language processing tasks.

2.2.5 Generative Transformers and Their Significance

Generative transformers have emerged as a groundbreaking advancement in the domain of artificial intelligence, particularly in natural language processing and generation. These models, characterized by their ability to generate coherent and contextually relevant sequences of text, have set new benchmarks in various tasks, from text completion to story generation. This section introduces the notable generative models available, including the GPT series and other significant contributions in this domain.

2.2.5.1 GPT (Generative Pre-trained Transformer) Series

The GPT series, developed by OpenAI, fully demonstrates the power and potential of generative transformers. Built upon the transformer architecture, the GPT models leverage the attention mechanism to understand and generate human-like text. The GPT series has seen rapid evolution, with each iteration bringing enhanced capabilities and performance.

GPT-1: The first in the series, GPT-1 [141], was released in 2018. It laid the foundation for subsequent models. With 117 million parameters, it showcased the potential of transformers in generating coherent paragraphs of text.

GPT-2: Released in 2019, GPT-2 [142] increased its parameters to 1.5 billion. Its ability to generate entire articles, answer questions, and even write poetry garnered significant attention from the research community and the public alike.

GPT-3: GPT-3 [17] has 175 billion parameters. Its capabilities extend beyond mere text generation; it can translate languages, write essays, create poetry, and even generate code.

GPT-4: The most recent model from OpenAI, GPT-4 [143], consists a staggering 1.76 trillion parameters, positioning it among the most advanced language models currently available. Leveraging advanced deep learning methodologies, it surpasses the capabilities of its forerunner, GPT-3. Remarkably, GPT-4 can handle up to 25,000 words simultaneously, a capacity eightfold greater than GPT-3. Furthermore, GPT-4 is versatile in accepting both text and image prompts, allowing users to define tasks across vision and language domains. A notable improvement in GPT-4 is its reduced propensity for hallucinations compared to earlier versions.

2.2.5.2 Other Notable Generative Transformer Models

Beyond the GPT series, the landscape of generative transformers is rich and diverse, with several models making significant contributions to the field.

BERT (Bidirectional Encoder Representations from Transformers): Developed by Google, BERT [111] revolutionized the way we approach natural language understanding tasks. Unlike GPT, which is generative, BERT is discriminative, designed to predict missing words in a sentence. Its bidirectional nature allows it to capture context from both the left and the right of a word, leading to superior performance in tasks like question-answering and sentiment analysis.

LLaMA: LLaMA [144] is an auto-regressive language model built on the transformer architecture, introduced by Meta. In February 2023, Meta unveiled the initial version of LLaMA, boasting 65 billion parameters adept at numerous generative AI functions. By July 2023, LLaMA 2 was launched with three distinct model sizes: 7, 13, and 70 billion parameters.

LaMDA: LaMDA [145] is a specialized family of transformer-based neural language models for dialog applications developed by Google in 2022. With up to 137 billion parameters and pre-training on 1.56 trillion words of public dialog and web text, LaMDA aims to address two key challenges: safety and factual grounding. The model incorporates fine-tuning and external knowledge consultation to improve its safety metrics, ensuring responses align with human values and avoid harmful or biased suggestions. For factual grounding, LaMDA employs external knowledge sources like information retrieval systems and calculators to generate responses that are not just plausible but also factually accurate. The model shows promise in various domains, including education and content recommendations, offering a balanced blend of quality, safety, and factual integrity.

2.3 Tutorial on Generative Transformers

In this section, we delve into a hands-on tutorial on generative transformers, guiding readers through the foundational concepts and practical implementations. By the end of this tutorial, readers should have a clear understanding of the transformer architecture and be equipped to build their own generative transformer models.

2.3.1 Basics of the Transformer Architecture

The transformer architecture, introduced by Vaswani et al. in their seminal paper "Attention Is All You Need" [109], has become the backbone of many state-of-the-art models in natural language processing. Let's break down its core components.

2.3.1.1 Overview

As depicted in Figure 2.3, the transformer consists of an encoder and a decoder. The encoder processes the input sequence, and the decoder generates the output sequence. Both the encoder and decoder are composed of multiple layers of attention mechanisms and feed-forward neural networks.

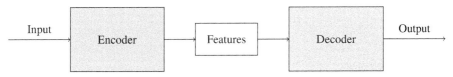

Figure 2.3: Expanded schematic representation of the transformer architecture with a smaller Features block.

2.3.1.2   Attention Mechanism

As previously discussed, the attention mechanism allows the model to focus on different parts of the input sequence when producing an output. The mechanism computes attention scores based on queries, keys, and values.

Mathematical Representation:
Given a query q, key k, and value v, the attention output is computed as:

$$\text{Attention}(q, k, v) = \text{softmax}\left(\frac{q \cdot k^T}{\sqrt{d_k}}\right) v \qquad (2.13)$$

where d_k is the dimension of the key.

Code Snippet:
The following Python code snippet demonstrates how to implement this attention mechanism using PyTorch:

```python
import torch
import torch.nn.functional as F

def scaled_dot_product_attention(q, k, v):
    matmul_qk = torch.matmul(q, k.transpose(-2, -1))
    d_k = q.size(-1) ** 0.5
    scaled_attention_logits = matmul_qk / d_k
    attention_weights = F.softmax(scaled_attention_logits,
        dim=-1)
    output = torch.matmul(attention_weights, v)
    return output, attention_weights
```

In this code snippet, q, k, and v are the query, key, and value tensors, respectively. The function *scaled_dot_product_attention* computes the attention output according to Equation 2.13.

2.3.1.3 Multi-Head Attention

Instead of using a single set of attention weights, the transformer uses multiple sets, allowing it to focus on different parts of the input simultaneously. This is known as multi-head attention.

Code Snippet:

```
class MultiHeadAttention(nn.Module):
    def __init__(self, d_model, num_heads):
        super(MultiHeadAttention, self).__init__()
        self.num_heads = num_heads
        # Dimension of the model
        self.d_model = d_model
        # Depth of each attention head
        self.depth = d_model
        # Linear layer for creating query, key and value
            matrix
        self.wq = nn.Linear(d_model, d_model)
        self.wk = nn.Linear(d_model, d_model)
        self.wv = nn.Linear(d_model, d_model)
        # Final linear layer to produce the output
        self.dense = nn.Linear(d_model, d_model)
```

2.3.1.4 Feed-Forward Neural Networks

Each transformer layer contains a feed-forward neural network, applied independently to each position.

Code Snippet:

```
class PointWiseFeedForwardNetwork(nn.Module):
    def __init__(self, d_model, dff):
        super(PointWiseFeedForwardNetwork,
            self).__init__()
        self.fc1 = nn.Linear(d_model, dff)
        self.fc2 = nn.Linear(dff, d_model)
        ...
```

Figure 2.4: PyTorch implementation of point-wise feed-forward network.

Each method and its body are indented with a tab or four spaces, which is the standard Python indentation. This makes the code easier to read and understand.

2.3.1.5 Self-attention Mechanism

The self-attention mechanism is a variant of the attention mechanism where the input sequence itself serves as the queries, keys, and values. This allows the transformer to weigh the significance of different parts of the input relative to a specific position, crucial for understanding context and semantics.

Mathematical Representation:
Given an input sequence X, the queries Q, keys K, and values V are derived as:

$$Q = XW_Q, \quad K = XW_K, \quad V = XW_V \tag{2.14}$$

where W_Q, W_K, and W_V are weight matrices. The self-attention output is then computed using the attention formula:

$$\text{SelfAttention}(Q, K, V) = \text{softmax}\left(\frac{QK^T}{\sqrt{d_k}}\right) V \tag{2.15}$$

2.3.1.6 Positional Encoding

Transformers, by design, do not have a built-in notion of sequence order. To provide the model with positional information, we inject positional encodings to the input embeddings. These encodings are added to the embeddings to ensure the model can make use of the sequence's order.

Mathematical Representation:
The positional encodings are computed using sine and cosine functions:

$$PE_{(pos,2i)} = \sin\left(\frac{pos}{10000^{2i/d_{\text{model}}}}\right) \tag{2.16}$$

$$PE_{(pos,2i+1)} = \cos\left(\frac{pos}{10000^{2i/d_{\text{model}}}}\right) \tag{2.17}$$

where *pos* is the position and i is the dimension.

2.3.1.7 Multi-head Attention

Multi-head attention is an extension of the attention mechanism, allowing the model to focus on different parts of the input simultaneously. By running multiple attention mechanisms in parallel, the model can capture various types of relationships in the data.

Mathematical Representation:

Given queries Q, keys K, and values V, the multi-head attention output is computed as:

$$\text{MultiHead}(Q,K,V) = \text{Concat}(\text{head}_1,\ldots,\text{head}_h)W_O \tag{2.18}$$

where each head is computed as:

$$\text{head}_i = \text{Attention}(QW_{Qi}, KW_{Ki}, VW_{Vi}) \tag{2.19}$$

and W_{Qi}, W_{Ki}, W_{Vi}, and W_O are weight matrices.

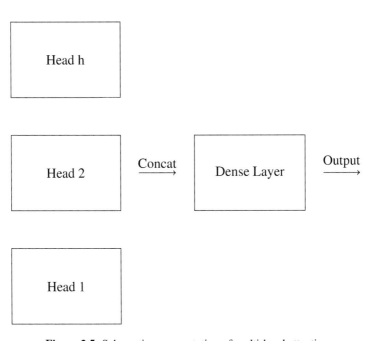

Figure 2.5: Schematic representation of multi-head attention.

Figure 2.5 showcases the multi-head attention mechanism, where multiple attention heads operate in parallel, and their outputs are concatenated and passed through a dense layer to produce the final output.

Understanding the intricacies of the transformer architecture, from the self-attention mechanism to multi-head attention, is crucial for harnessing its full potential. By delving into the mathematical foundations and practical implementations, one can build powerful models capable of handling a wide range of tasks in natural language processing.

2.3.1.8   Encoder and Decoder Modules

The Transformer architecture consists of an encoder and a decoder, each made up of multiple layers. Here, we'll walk through the implementation of these modules.

Encoder Module:
The encoder module consists of multiple encoder layers, each containing multi-head attention and feed-forward neural networks.

Code Snippet for Encoder:

```
import torch.nn as nn

class EncoderLayer(nn.Module):
    def __init__(self, d_model, num_heads):
        super(EncoderLayer, self).__init__()
        self.mha = MultiHeadAttention(d_model,
            num_heads)
        self.ffn =
            PointWiseFeedForwardNetwork(d_model, dff)
        # Layer normalization and dropout layers can
            be added here

    def forward(self, x):
        attn_output = self.mha(x, x, x)
        out1 = x + attn_output   # Add & Norm
        ffn_output = self.ffn(out1)
        out2 = out1 + ffn_output   # Add & Norm
        return out2
```

Decoder Module:
The decoder module is similar to the encoder but has an additional multi-head attention layer to attend to the encoder's output.

Code Snippet for Decoder:

```
class DecoderLayer(nn.Module):
    def __init__(self, d_model, num_heads):
        super(DecoderLayer, self).__init__()
        self.mha1 = MultiHeadAttention(d_model,
            num_heads)
        self.mha2 = MultiHeadAttention(d_model,
            num_heads)
        self.ffn =
            PointWiseFeedForwardNetwork(d_model, dff)
        # Layer normalization and dropout layers can
            be added here
```

```
def forward(self, x, enc_output):
    attn1 = self.mha1(x, x, x)
    out1 = x + attn1    # Add & Norm
    attn2 = self.mha2(out1, enc_output, enc_output)
    out2 = out1 + attn2    # Add & Norm
    ffn_output = self.ffn(out2)
    out3 = out2 + ffn_output    # Add & Norm
    return out3
```

In these code snippets, 'MultiHeadAttention' and 'PointWiseFeedForward-Network' are custom classes that you would define based on your specific needs for multi-head attention and point-wise feed-forward networks, respectively.

2.3.2 Building a Simple Generative Transformer

Building a generative transformer from scratch involves several steps, from data preprocessing to model training and text generation. In this section, we'll walk through each of these steps, providing a comprehensive guide to constructing your own generative transformer.

2.3.2.1 Data Preprocessing and Tokenization

Before feeding data into the model, it's essential to preprocess and tokenize it. Tokenization involves converting raw text into a sequence of tokens, which can be words, subwords, or characters.

Tokenization:
Using popular libraries like the HuggingFace's 'transformers', tokenization can be achieved as:

```
from transformers import GPT2Tokenizer

tokenizer =
    GPT2Tokenizer.from_pretrained('gpt2-medium')
tokens = tokenizer.encode("Hello,␣world!")
```

2.3.2.2 Defining the Transformer Model

Assuming you've already defined the EncoderLayer and DecoderLayer classes, You can define the complete Transformer model as follows:

```
class Transformer(nn.Module):
    def __init__(self, d_model, num_heads, num_layers):
        super(Transformer, self).__init__()
        self.encoder =
            nn.ModuleList([EncoderLayer(d_model,
            num_heads) for _ in range(num_layers)])
```

```
        self.decoder =
            nn.ModuleList([DecoderLayer(d_model,
            num_heads) for _ in range(num_layers)])

    def forward(self, src, tgt):
        enc_output = src
        for layer in self.encoder:
            enc_output = layer(enc_output)

        dec_output = tgt
        for layer in self.decoder:
            dec_output = layer(dec_output, enc_output)

        return dec_output
```

Building a generative transformer, while complex, is made accessible with modern libraries and tools. By understanding the steps involved, from data preprocessing to model training and generation, one can harness the power of transformers for a wide range of applications.

2.3.3 *Advanced Techniques and Best Practices*

While the foundational concepts and basic implementations provide a solid starting point, mastering generative transformers requires a deeper understanding of advanced techniques and best practices. This section offers insights into improving generation quality, handling long sequences, memory issues, and leveraging fine-tuning and transfer learning [146].

2.3.3.1 *Techniques for Improving Generation Quality*

Achieving high-quality text generation necessitates a combination of model architecture tweaks, training strategies, and post-processing methods.

Temperature Sampling:
By adjusting the temperature during sampling, one can control the randomness of the generated text [147]. A lower temperature makes the output more deterministic, while a higher value introduces randomness.

$$p_i = \frac{e^{\frac{z_i}{T}}}{\sum_j e^{\frac{z_j}{T}}} \tag{2.20}$$

where p_i is the adjusted probability, z_i is the original probability, and T is the temperature.

Top-k and Top-p Sampling:
Instead of sampling from the entire distribution, one can restrict the sampling pool to the top-k tokens or those tokens that have a cumulative probability greater than a threshold p [148].

Gradient Clipping:
To prevent exploding gradients during training, gradient clipping can be employed, ensuring the gradients remain within a defined range [149]. Gradient clipping can be implemented in PyTorch as follows:

```
torch.nn.utils.clip_grad_norm_(model.parameters(),
    max_norm=1.0)
```

2.3.3.2 Handling Long Sequences and Memory Issues

Transformers, by design, have quadratic complexity with respect to sequence length. This can lead to memory issues for long sequences.

Gradient Accumulation:
Instead of updating the model weights after every batch, gradients can be accumulated over multiple batches, effectively simulating a larger batch size without the memory overhead [150].

Model Parallelism:
For models with billions of parameters, distributing the model across multiple GPUs can alleviate memory constraints [151].

Gradient Checkpointing:
This technique involves storing intermediate activations during the forward pass and recomputing them during the backward pass, reducing memory usage at the cost of increased computation.

2.3.3.3 Fine-tuning and Transfer Learning

Transfer learning, the practice of leveraging pre-trained models on new tasks, has proven highly effective in the NLP domain.

Fine-tuning:
Once a model is pre-trained on a large corpus, it can be fine-tuned on a smaller, task-specific dataset. This approach often yields superior results compared to training from scratch [152, 153].

Adapters:
Instead of fine-tuning the entire model, adapters allow for training only a small portion of the model, introducing task-specific parameters without altering the pre-trained weights [154].

Mastering generative transformers goes beyond understanding the basics. By incorporating advanced techniques and best practices, one can achieve state-of-the-art performance, handle large models and sequences efficiently, and adapt pre-trained models to new tasks with ease. As the field of NLP continues to evolve, staying abreast of these practices ensures robust and high-quality model deployments.

2.4 Applications and Use Cases

Generative transformers, with their unparalleled capability to understand and generate human-like text, have found applications across a myriad of domains [120]. This section provides an in-depth exploration of some of the most prominent applications, shedding light on the transformative impact of these models on various industries.

2.4.1 Text Generation for Creative Writing

The realm of creative writing, traditionally seen as the bastion of human creativity, has witnessed significant advancements with the advent of generative transformers [155]. These models, trained on vast corpora of literature, can produce text that mirrors the style, tone, and complexity of human authors.

Novel and Short Story Generation: AI-powered applications based on GPT-3 and other large language models have been employed to generate entire novels or assist authors by suggesting plot twists, character developments, and dialogues [156]. The generated content, while sometimes requiring human oversight, exhibits creativity and coherence.

Poetry and Song Lyrics: The nuanced and abstract nature of poetry and song lyrics poses a significant challenge for traditional models. However, the advent of generative transformers has enabled these models to produce verses that resonate with human emotions and experiences. A recent study demonstrated that AI-generated poems were often indistinguishable from those written by humans [157], showcasing the success of these algorithms in replicating human-like poetic expressions.

2.4.2 Chatbots and Conversational Agents

The rise of digital communication has spurred the demand for intelligent chatbots and conversational agents. Generative transformers, with their ability to generate contextually relevant and coherent responses, stand at the forefront of this revolution. One of the most prominent examples of a conversational agent built on generative transformer architecture is ChatGPT, developed by OpenAI. Chat-GPT reached 100 million monthly active users just two months after launching, making it the fastest-growing application in history.

Customer Support: Businesses employ transformer-based chatbots to handle customer queries, complaints, and feedback [158, 159]. These chatbots can understand the context, provide accurate information, and even escalate issues when necessary.

Personal Assistants: Digital personal assistants, like Siri and Alexa, are integrating transformer models to enhance their conversational capabilities, making interactions more natural and context-aware.

2.4.3   Code Generation and Programming Assistance

Software development is undergoing a significant transformation with the introduction of transformer models capable of understanding and generating code. One such model that transforms natural language instructions to code is the Codex model developed by OpenAI [160]. These models assist developers by suggesting code snippets, detecting bugs, and even generating entire functions or modules.

Code Completion: Integrated Development Environments (IDEs) are incorporating transformers to provide real-time code completion suggestions, enhancing developer productivity.

Bug Detection and Fixing: Transformers can be trained to detect anomalies in code and suggest potential fixes, reducing debugging time and ensuring more robust software.

2.4.4   Other Notable Applications

Beyond the aforementioned domains, generative transformers have found applications in diverse areas:

Translation: While traditional machine translation models have limitations, transformers can produce translations that consider the broader context, resulting in more accurate and idiomatic outputs [114].

Summarization: Generative transformers can read lengthy articles or documents and produce concise summaries, retaining the core information and intent [115].

Gaming: In the gaming industry, transformers are used to generate dialogues, plotlines, and even assist in game design by suggesting scenarios or character backstories [161].

The applications of generative transformers are vast and continually expanding. As research progresses and models become more sophisticated, it is anticipated that their integration into various domains will become even more profound.

2.5 Challenges and Limitations

While generative transformers have showcased remarkable capabilities, they are not devoid of challenges and limitations. This section delves into some of the most pressing concerns surrounding these models, from interpretability issues to ethical dilemmas and computational constraints.

2.5.1 Model Interpretability

Deep learning models, especially those with millions or billions of parameters like generative transformers, are often criticized for being "black boxes." Understanding why a model made a particular decision can be elusive [162].

Attention Maps: One approach to interpretability is visualizing attention maps [109, 163]. These maps show which parts of the input the model focused on when producing an output. Attention maps are generated by the attention mechanism that computes a set of attention scores, which can be visualized as a heat map.

Attention maps serve as a tool for interpreting transformer models in NLP by providing insights into various aspects of text processing. They help in analyzing the roles of words in sentences, identifying key topics, evaluating text quality, and detecting errors or biases. However, while attention maps provide insights, they don't offer a complete understanding of the model's decision-making process.

Mathematical Analysis: Efforts are being made to develop mathematical tools and frameworks to dissect the inner workings of transformers [164, 165]. Yet, a comprehensive understanding remains a research frontier.

2.5.2 Hallucination in Text Generation

Generative transformers are sometimes susceptible to generating text that, while coherent and grammatically correct, is factually incorrect or nonsensical. This phenomenon is commonly referred to as hallucination. Ji et al. conducted a comprehensive survey of the issue of hallucination in natural language generation (NLG) [166].

The causes of hallucination are multifaceted and can vary. They may include inadequate training data, which limits the model's understanding of the subject matter. Overfitting to the training set is another common issue, where the model learns the noise in the data rather than the actual pattern. Additionally, high model complexity leading to over-parameterization can also contribute to hallucination.

Addressing the issue of hallucination involves multiple strategies. One approach is to fine-tune the model on a more specific dataset that is closely aligned with the task at hand. Another strategy involves incorporating external knowledge bases that can fact-check the generated text in real-time. Ensemble methods, which combine the outputs of multiple models, can also be used to validate the generated text and reduce the likelihood of hallucination.

Efforts are underway to quantify the degree of hallucination in generated text. Although a standard measure has yet to be established, one simplistic way to quantify it is through the Hallucination Score, defined as the ratio of the number of hallucinated tokens to the total number of generated tokens, as shown in Equation 2.21.

$$\text{Hallucination Score} = \frac{\text{Number of hallucinated tokens}}{\text{Total number of generated tokens}} \quad (2.21)$$

2.5.3 Ethical Considerations in Text Generation

Generative transformers, with their ability to produce human-like text, raise several ethical concerns [167].

Misinformation and Fake News: There's potential for these models to generate misleading or false information, which can be weaponized to spread misinformation.

Bias and Fairness: Transformers, being trained on vast internet datasets, can inherit and perpetuate biases present in the data [168]. Addressing this requires careful dataset curation and post-hoc bias mitigation techniques.

$$\text{Bias} = \frac{\sum_{i=1}^{n} (P_{\text{model}}(x_i) - P_{\text{true}}(x_i))}{n} \quad (2.22)$$

where P_{model} is the model's prediction, P_{true} is the true distribution, and n is the number of samples.

2.5.4 Computational Requirements and Environmental Impact

Training a large language model demands significant computational resources. For example, the GPT-3 model with 175 billion parameters would require $3.14e^{23}$ FLOPS for training, translating to 355 GPU-years and a cost of \$4.6 million on a V100 GPU [169]. Memory is another bottleneck; the model's 175 billion parameters would need 700GB of memory, far exceeding the capacity of a single GPU. To manage these challenges, OpenAI used model parallelism techniques and trained the models on a high-bandwidth cluster. As language models grow in size, model parallelism is becoming increasingly essential for research.

Energy Consumption: The energy required to train state-of-the-art models can be equivalent to the carbon footprint of multiple car lifetimes. This raises environmental concerns.

Exclusivity: The computational demands mean that only well-funded organizations can train the most advanced models, leading to concerns about the democratization of AI.

While generative transformers offer immense potential, it's crucial to address their challenges and limitations. Balancing the pursuit of state-of-the-art performance with ethical, environmental, and computational considerations is paramount for the sustainable and responsible advancement of the field.

2.6 The Future of Generative Transformers

Generative transformers, evolving from early models like the Recurrent Neural Networks (RNNs) to the sophisticated Generative Adversarial Networks (GANs) and now the powerful transformers, have revolutionized numerous domains. With advancements in model architectures, training techniques, and hardware capabilities, we can anticipate models that not only understand and generate human-like text but also exhibit enhanced creativity, reasoning, and a form of artificial consciousness.

The way forward is full of opportunities for exploration and innovation. As the field of generative transformers continues to evolve, there are numerous avenues for research and development that remain unexplored or underexplored. The evolution from rules-based systems to advanced LLMs has dramatically improved performance and training efficiency. These improvements are not confined to text and language processing but extend to computer vision and other modalities, creating avenues for interdisciplinary research.

2.6.1 Multimodal models

The future sees generative models that seamlessly integrate multiple modalities – text, image, sound, video, and more – offering a holistic understanding of the world and generating content that overcomes the limitations of current models. Recent advancements have already led to transformers capable of generating not just text, but also image, audio, and video [170]. These multimodal models are expected to evolve into sophisticated systems capable of processing and understanding inputs from various modalities simultaneously.

In the future, we anticipate the emergence of single applications and more advanced multimodal models. These systems would not only understand inputs from different sensory channels - such as visual, auditory, and textual - but also generate outputs in various forms, moving well beyond mere text generation. The integration of these modalities in a single model offers a more comprehensive approach to understanding complex real-world scenarios and creating more nuanced and contextually relevant outputs.

2.6.2 Domain-Specific Models

The development of domain-specific GPT models is becoming increasingly crucial across various applications [171]. While current large language models are adept at understanding natural language and generating content, their effectiveness and accuracy can vary significantly when applied to specialized domains such as medicine, law, and finance [172]. A big challenge in tailoring these models to a specific domain lies in the acquisition of high-quality, domain-specific data. Another significant challenge is the fine-tuning process, which involves adapting the model to the unique characteristics and vocabulary of the domain.

Despite these obstacles, there has been progress in the development and implementation of domain-specific GPT models. The emergence of these models marks a future towards more tailored AI solutions. Companies with unique large datasets stand to gain competitive advantages by training their own bespoke models. This trend is exemplified by Bloomberg's development of a specialized LLM for financial tasks [173]. Other ompanies like Hugging Face and Databricks are also playing pivotal roles in providing the necessary resources and platforms for developing and fine-tuning these customized models.

In the future, we can expect these domain-specific GPT models to offer enhanced efficiency, improved interpretability, and better domain generability compared to existing large language models. However, the development of these models must also focus on optimizing energy consumption and addressing the challenges of knowledge retention during the fine-tuning process.

2.6.3 Model Efficiency

The growing size of models necessitates research in computational efficiency and energy consumption. This includes efforts to develop more sustainable AI infrastructure and predictive infrastructure, essential for the data-intensive nature of enterprise AI applications.

2.6.4 Ethical AI

With the widespread implementation of generative AI across various sectors, ensuring ethical use becomes paramount. This involves research into bias mitigation, fairness, transparency, and the development of guidelines for responsible AI usage [174], especially as AI begins to automate complex tasks like legal work and medical fields like drug design and medical diagnosis.

2.6.5 Interdisciplinary Integration

The future of generative AI involves its fusion with other fields like neuroscience and cognitive science. This integration could lead to breakthroughs in understanding both artificial and natural intelligence, with generative AI applications expanding beyond technical fields to impact popular culture and everyday life, such as in the creation of high-resolution images and user-friendly AI applications for enhancing productivity.

As we reflect upon the evolution of generative transformers, from their foundational roots with Alan Turing to their current state-of-the-art capabilities, it becomes clear that we are at a turning point in the development of artificial intelligence. In the words of Alan Turing, "We can only see a short distance ahead, but we can see plenty there that needs to be done."

2.7 Conclusion

As we reflect upon the evolution of generative transformers, from their foundational roots with Alan Turing to their current state-of-the-art capabilities, it becomes clear that we are at a turning point in the development of artificial intelligence. In the words of Alan Turing, "We can only see a short distance ahead, but we can see plenty there that needs to be done." This foresight aptly describes the current state of AI. The advancements in generative transformers have not only redefined what machines are capable of doing but also opened up a myriad of possibilities for future exploration and innovation. As we advance and develop new technologies, it is crucial to navigate the ethical implications, environmental and societal impacts of these technologies. The goal is not just to push the boundaries of what AI can achieve but to do so responsibly, ensuring that these advancements benefit society at large.

Chapter 3

Ethical Considerations for AI

3.1 Bias and Fairness - Ethical Challenges of Biased Algorithms and Decision-making

3.1.1 Introduction

3.1.1.1 Background

In recent decades, algorithms and artificial intelligence (AI) systems have increasingly been deployed across many high-impact domains to automate complex decision-making processes. From approving loan applications to predicting recidivism risks in the criminal justice system, AI and machine learning models are now deeply embedded in processes that significantly shape human lives [175]. The exponential growth in computational power has enabled this rapid adoption, the rise of big data, and breakthroughs in machine learning techniques like deep neural networks. However, handing over sensitive decisions to opaque algorithmic systems has also sparked widespread concerns about unfairness, accountability, and unintended consequences [43]. This is an overview of the historical rise of automated decision-making via algorithms and AI, analyzing key technical developments and applications while facing ethical tensions and challenges.

The vision of developing systems that can augment or replicate human decision-making dates back decades before the era of modern AI. In the 1940s and 1950s, pioneering researchers like Alan Turing conceptualized intelligent machines, laying the theoretical foundations for artificial intelligence [176]. The earliest automated decision systems relied on simple rule-based approaches with predefined logic and criteria. For instance, credit scoring systems introduced in the 1950s helped lenders automate loan approval decisions by ranking applicants against financial criteria and cutoffs [177]. However, these systems required explicit programming of decision rules by domain experts. They remained relatively narrow in scope of application.

Starting in the 1980s, machine learning techniques allowed computers to derive decision-making models directly from data. Rather than relying on predefined rules, algorithms could extract patterns from training data to make predictions and classifications [178]. This enabled automated decision systems that evolved and improved with more data without explicit programming. For example, machine learning enabled early fraud detection systems in banking to model users' behaviour and flag anomalies [179]. Adoption significantly accelerated in the 2010s with the rise of deep learning, fuelled by growth in computing power and the availability of massive datasets for training. Deep neural networks could recognize patterns in highly complex, unstructured data like images, video and text. This vastly expanded the applicability of AI to diverse decision domains.

Today, AI decision systems utilize vast data and advanced algorithms to match or exceed human performance on many specialized tasks. In contexts like medical diagnosis and self-driving vehicles, AI is lauded for its potential to save lives by making faster, more accurate decisions than humans [180]. However, critics highlight that incorporating AI into high-stakes social systems introduces significant ethical risks. The complexity of deep learning models makes their internal decision-making processes opaque, lacking interpretability and explainability [181]. This becomes concerning when AI is deployed in sensitive contexts like criminal justice, where biased predictions or unfair outcomes could result without accountability. Understanding the historical evolution of automated decision-making systems using AI and data is crucial to grapple with societal impacts. To concretely illustrate the proliferation of AI decision systems, it is instructive to survey some major domains of application and ethical tensions that have emerged:

1. Criminal Justice: AI tools are used widely by law enforcement and courts to predict crime hotspots, assess re-offending risks, and recommend bail or parole decisions [182]. These predictive systems have faced scrutiny for potential racial biases and lack of due process.

2. Healthcare: AI shows promise to aid clinical decision-making in areas like medical imaging, patient triage and diagnosis [180]. However, concerns exist about over-reliance on AI predictions, accountability gaps, and patient consent.

3. Hiring: Automated resume screening and candidate assessment based on data mining and machine learning are now standard in recruitment [183]. Critics argue that this can perpetuate biases and reduce diverse hiring.

4. Finance: AI is deployed extensively in banking for functions like credit lending decisions, insurance underwriting, and stock trading algorithmic trading. However, it can replicate historical biases against marginalized groups [42]. These examples demonstrate that AI decision systems are transforming many industries and raising pressing ethical questions regarding fairness, accountability, and transparency.

One fundamental concern is that training datasets used in machine learning contain ingrained human biases reflecting long legacies of discrimination and injustice in societies [184]. Models derive their decision rules purely from spotting statistical patterns in data. Without careful oversight, they readily pick up and perpetuate biases like race or gender prejudice from the past. For instance, natural language processing models consistently acquire gender stereotypes, like associating science careers with men simply by analyzing patterns in text corpora from the wider culture [185]. Biased data reflects complex historical forces which do not arise from a few explicitly prejudiced labels. For example, recruiter assessments in job training data contain layers of societal biases. Credit lending decisions reflect decades of deliberately racist policies that continue limiting opportunities for people of colour. AI models do not comprehend the unjust social origins of learned patterns - they replicate and entrench discriminatory logic [186]. This risks automating and further normalizing long legacies of injustice. The complex, technical nature of algorithms also makes discrimination harder to detect. Whereas human decision-making may explicitly invoke unacceptable criteria like race, opaque AI systems can achieve discriminatory ends implicitly through technical means that conceal unfairness. For instance, machine learning models can latch onto non-intuitive proxies correlated with race, like zip codes or purchase patterns, to exclude minorities without mentioning race directly [187]. The sheer complexity of deep learning techniques makes it nearly impossible to thoroughly inspect models' reasoning and logic. This obfuscation allows potentially objectionable practices to avoid scrutiny behind a veneer of technical neutrality and objectivity [43].

Some argue that this technical mediation could make discrimination more insidious, as people blame unfair outcomes on algorithms rather than take responsibility for conscious prejudice [188]. Overall, the non-intuitive nature of bias encoded in opaque models poses significant accountability challenges. It becomes difficult to fully understand why some groups face consistent disadvantages or assess if unfair reasoning underlies decisions [189]. This clashes with principles of equal treatment and non-discrimination. The complexity of modern AI techniques also often makes it impossible to explain or audit their reasoning processes. For example, deep neural networks derive patterns by transforming input data across thousands of nodes in hidden layers according to non-linear functions - even developers struggle to articulate precisely how outputs are produced [189]. This opacity becomes ethically problematic when life-changing decisions are affected. A denied loan, prison sentence or other serious impact demands recourse to inspect unfair treatment and contest decisions. However, opaque AI systems lack mechanisms for impacted individuals to understand outcomes or dispute potential errors [190]. This unaccountability clashes with norms of procedural justice, preventing subjects from participating in decisions affecting them. It also undermines due process principles and the right to appeal [191]. Discrimination and mistakes can no longer be systematically detected or corrected. While algorithms are often considered neutral and objective, their opacity gravely limits how unfairness and inaccuracy can be recognized or redressed. This allows injustice to persist unchecked behind the technology.

Finally, integrating algorithmic systems changes decision processes in ways that can circumvent established functions for promoting fairness. For example, replacing human judgement steps with automated models in business hiring pipelines disrupts mechanisms to discern and mitigate bias, such as diversity committees reviewing decisions to flag possible prejudice [183]. New algorithmic mediation alters pipelines to prevent existing quality control points designed to uphold ethics. Likewise, incorporating risk scores into criminal justice processes like bail and sentencing prevents judges from considering the individual circumstances or societal disadvantages that shaped an offender's trajectory. This overrides a crucial role of human discretion in attempting to enact justice, not just legalistic outcomes [192]. Automated, data-driven decisions can achieve efficiency gains by shedding nuanced functions evolved to promote fairness and prevent discrimination. However, this clashes with society's moral responsibility to remedy bias and structural disadvantage.

3.1.1.2   Sources and Nature of Algorithmic Bias

A significant source of bias in artificial intelligence (AI) systems is the historical biases in the training data used to develop algorithms and machine learning models. Societal discrimination and injustices accumulated over decades or centuries become encoded into datasets, reflecting long legacies of exclusion or prejudice against marginalized groups [184]. Here, the discussion is focused on how deeply rooted societal biases get reflected in data, analyzing their historical origins, complexity, and propagation into AI systems with risky impacts on fairness and justice. It is crucial to examine history to understand how data comes to reflect societal bias. Data is generated through social institutions and practices shaped by centuries of politics, culture, economics and power dynamics. These forces imprint biased assumptions and unfair prejudices within the records, statistics and documents, which are then used as data. For instance, policing data encoding racial profiling reflects decades of discriminatory policy and culture in law enforcement [186]. The healthcare systems entrenched with racist attitudes and inequities produce medical datasets with underdiagnoses for minority patients. Such biased data thus has complex historical roots spanning generations, and correcting such multifaceted generational biases requires reckoning with complicated histories. Once discrimination becomes encoded into datasets, this inherits into any systems later built from that data. For example, facial recognition datasets disproportionately containing white faces reflect histories of exclusion in photography, media and technology. However, this imbalance gets propagated as new AI models rely on the same data [185]. Often, dominant groups already overrepresented in past data gain a further advantage by having more training data devoted to improving systems' performance for them. Without conscious countermeasures, algorithms easily amplify inequality as data circulates between institutions.

Leaving such inherited bias unacknowledged allows history to perpetuate injustice through data indefinitely. However, many technical practitioners view data as objective ground truth rather than as a socially constructed artefact [193]. This data determinism assumes that training datasets neutrally represent reality rather than mirror long-distorted social systems. Inheritance effects make distinguishing where current biases originated and replicated in new contexts difficult. However, not tracing such lineages helps institutions evade responsibility for disparate outcomes. In addition to inherent prejudice in the data itself, the data collection and labelling process often adds further bias, even absent ill intent. When humans manually annotate datasets, unconscious biases influence what information gets recorded, how it gets categorized, and who gets represented in the data acquisition process. Social psychology reveals many ways unconscious cognitive biases related to race, gender, appearance and cultural background distort human judgement in subjective tasks like annotation [194]. For example, microscope image datasets for AI diagnosis systems were predominantly annotated by male physicians, contributing to worse performance on images of female patients [195]. Such implicit biases easily propagate into AI, given the lack of diversity among data labellers and dependence on narrow subjective viewpoints. Even when datasets derive directly from sensors rather than human input, what gets captured reflects social choices and prioritization. Overall, a broad range of human decisions in generating data adds implicit layers of bias atop inherited prejudice.

Feeding historically biased data into AI systems compounds the risks of perpetuating injustice. Machine learning models mathematically amplify patterns, entrenching whatever biases exist in data. Algorithms also lack capacities for social context and ethical reasoning that humans can employ to counter prejudice. Nuance, empathy and systemic perspectives are inherently difficult for AI [193]. As a result, AI systems readily pick up and exacerbate even subtle biases, leading to discriminatory impacts. Without understanding how datasets accumulate historical prejudices, it becomes impossible to intervene and prevent harm to fairness. If institutions correct future data collection to be more inclusive, new issues arise in combining this with old biased data. Attempting to utilize all available training data could spread outdated biases into newer and fairer datasets. However, discarding older data also erases historically marginalized groups from current models. There are no easy solutions once bias takes root and circulates through data dependencies and reuse. Responsible approaches require grappling with complex histories of discrimination.

3.1.1.3 Lack of Diversity and Technical Limitations

In addition to historical data biases, characteristics of the technology sector, like lack of diversity and technical constraints, also raise risks of bias during the development process of artificial intelligence (AI) systems and algorithms. Homogenous teams building AI can introduce groupthink and blindspots, while technical complexities make auditing systems and addressing flaws difficult. This section will analyse how limitations in the makeup of AI development teams and technical AI methods contribute to potential biases and ethical pitfalls in algorithm design. Currently, those building and deploying AI systems predominantly comprise privileged demographics. The technology sector grapples with immense gender gaps, with women holding just 25% of computing roles at major tech firms as of 2021 [196]. Black and Latinx groups remain drastically underrepresented, constituting just 5% and 3% of Silicon Valley's workforce, respectively. Engineers hail overwhelmingly from a narrow Western, educated background. This lack of perspective diversity has significant implications for biases that enter AI design. Product teams lacking underrepresented voices are less attuned to the concerns or needs of marginalized groups. Unconscious biases shared among homogenous developers are likelier to go unquestioned, leading AI to amplify social prejudices. For instance, facial recognition systems have systematically performed worse on women and darker skin types, reflecting the historical dominance of pale male engineers in the field [197]. Diversity is vital for catching blindspots.

The importance of representation spans not just developers but also data annotators, subject reviewers for testing, and deployment contexts. Skewed participant pools during training data collection or trials can hide failures for underrepresented populations before real-world deployment. Overall, AI built inside tech's homogeneous culture risks harming those excluded from the design process. Experts often cannot explain precisely why an AI model derived a certain prediction or decision [198]. The system's emergent reasoning remains partly concealed within a black box model too chaotic to fully inspect or understand, let alone predict flaws before usage. While some interpretability methods seek to shed light on these black boxes, they offer incomplete solutions [181]. Overall, the inherent complexity of modern AI techniques poses barriers to auditing systems robustly for issues like bias or fairness. These technical constraints also cause challenges when attempting to correct issues identified in algorithms after initial deployment. Given system complexity, problems detected in use, like gender or racial biases, often cannot be easily isolated to root causes for repair. Unlike code, there is no intuitive way to edit discrete logic in neural networks directly. Fixes instead often rely on indirect tweaks like seeking to balance biased training data or adding constraints to output variables [184]. However, these approaches risk unintended effects that create new problems elsewhere in the model. There are also challenges in how modifications will transfer. Updating a model on a device may be infeasible, requiring propagation across distributed copies. Limited interpretability makes it hard to verify changes. Overall, it remains an open research question in AI how to efficiently debug and correct issues in complex models once created - unlike software code, logic in models remains diffuse. These uncertainties pose risks of deploying algorithms with unresolved fairness flaws. The lack of assurance in correcting issues is compounded as technology companies deploy AI systems at massive scales. Platforms like Facebook and Google operate prediction algorithms impacting billions of users. However, their closed development environments pose significant accountability challenges for bias. Their models and data remain mostly proprietary black boxes impenetrable to external auditing or peer review. Reliance on internal cultures of ethics is concerning, given the lack of workforce diversity [197]. The systems' scale also means flawed logic could cause harm worldwide before issues are flagged. There are inadequate incentives for companies to sacrifice business interests like engagement for bias corrections that reduce profitability. The technical complexity of modern AI paired with industrial deployment poses systemic risks of algorithmic bias and discrimination going unchecked on massive scales.

3.1.1.4 Blindspots and Assumptions in Model Development

In building artificial intelligence (AI) systems, developers make many choices and assumptions that shape how algorithms function during the model development process. However, blindspots, cognitive biases, and a lack of diverse perspectives often influence these decisions. As a result, problematic assumptions can get designed unconsciously into AI models, leading to issues like unfair biases and discrimination emerging in real-world usage. Let us analyze common pitfalls during model development that introduce bias risks and discuss strategies to improve awareness of assumptions in algorithm design.

Problem framing - formulating the task an AI model aims to perform is the first area where limited perspectives lead developers astray. Formulating tasks often relies on subjective human judgement of what variables and data are relevant. Without considering wider social contexts, developers commonly operationalize problems through narrow statistical measures. However, this can build unfair assumptions and blindspots into models when issues arise from complex structures like racism that pure data science cannot easily represent. Ethical AI requires sincerely grappling with full socio-technical contexts, not just technical variables, in isolation [199]. Many default assumptions go unexamined in collecting the training data to develop models, shaping what information algorithms learn. Often, convenience and availability primarily dictate data sources without evaluating whether datasets sufficiently cover impacted populations or feature essential variables. For example, facial analysis datasets constructed by simply scraping celebrity photos online create risks of bias, given the lack of diversity and consent [200].

However, vague acquiescence to notions like "more data is better" allows issues in unrepresentative datasets to remain unchallenged. Problematically, marginalized groups often have far less access to resources, allowing their participation in data collection. However, when left out of datasets, their needs get excluded from model development. Subjecting data collection practices to ethical scrutiny is vital to avoid encoding unjust assumptions into AI. Narrow assumptions also frequently arise in how model performance gets evaluated and which optimization objectives algorithms pursue. While metrics like accuracy seem intuitive, optimizing blindly for accuracy risks overserving majority groups. For example, always recommending the most common products satisfies accuracy but overlooks minority preferences [201]. Problematically, marginalized populations often comprise outliers in data. Optimizing only mainstream accuracy leads algorithms to dismiss their needs and encode bias. However, developers often apply such metrics instinctively without examining alignment with values like fairness. Rethinking evaluation requires grappling with difficult questions of which groups deserve representation and how to quantify the harms of underserving different users. Nevertheless, confronting these dilemmas is imperative to avoid baking prejudice into supposedly neutral metrics. A blindspot arises in failing to consider model impacts across different social groups during development and testing. AI systems are often trialled only on mainstream populations from which training data originated. Models propagate to actual usage without proactively including disadvantaged identities in testing pools with unknown harms towards marginalized users [202]. For instance, speech recognition performs vastly worse for deaf users in the absence of deliberate efforts to address this - an enormous oversight. Mostly, developers instinctively overlook minority groups and edge cases when assessing models, focusing only on common users. However, technology premised on indifference to marginalized populations can cause grave collective harm even without ill intent by any designer. Alleviating this requires intentional practices to evaluate models across diverse demographics and prevent excluding users invisible to developers.

3.1.2 The Impacts of Algorithmic Bias

3.1.2.1 Discrimination and Unjust Outcomes for Marginalized Groups

One of the most concerning impacts of bias in artificial intelligence (AI) systems is that it can lead to discrimination and unjust outcomes against already marginalized groups in society. There are many documented cases across crucial domains like employment, healthcare, and criminal justice where biased algorithms have created or exacerbated discrimination [203]. This compounds existing structural inequalities harming vulnerable populations. This section will present research on algorithmic discrimination across critical contexts, analyzing the discriminatory impacts on minority groups and violations of principles of justice and fairness. AI-driven hiring and recruitment tools have rapidly proliferated in recent years. However, research reveals consistent issues of gender and racial bias leading to discrimination. Amazon scrapped an algorithmic resume screening tool after discovering it discriminated against women by penalizing resumes containing the word "women's" [204]. Such algorithmic biases compound long legacies of employment discrimination, forcing underrepresented groups to navigate unjust barriers to thrive in workplaces. Biased AI decision-making violates equal access and treatment principles, contravening anti-discrimination employment laws.

In healthcare, numerous studies have revealed issues of racial, gender, and age bias in AI tools, leading to misdiagnoses or substandard treatment recommendations that exacerbate existing inequities. An algorithm to triage patients for extra medical care disadvantaged Black patients by underestimating their needs relative to equally sick white patients [203]. Medical AI may rationalise unequal treatment as an objective fact derived from data. However, all individuals deserve equal access to quality healthcare independent of identity. Removing biased assumptions and foregrounding this ethical commitment in medical AI development is imperative. The criminal justice domain provides some of the most troubling examples of discriminatory AI. Racially biased predictive policing tools impacted minorities disproportionately. Algorithms predicting re-offending risk scores discriminate against Black defendants, denying them equal parole opportunities [192]. Criminal justice AI operates on highly biased data reflecting decades of racist policies and prejudice. Systems claiming to forecast crime objectively reproduce past patterns of racial profiling, demonstrating how algorithmic bias causes grave harm to justice, liberty and human rights when applied to areas with an extensive history of discrimination.

3.1.2.2 Perpetuating and Exacerbating Wider Social Inequities

Beyond direct discrimination, biased algorithms also create broader harms by entrenching and worsening structural inequalities that already marginalize certain groups in society. Even absent overt prejudice in any one system, AI can act as a regressive force that widens disparities and worsens outcomes for vulnerable populations when deployed uncritically within unjust social contexts. This section will examine scholarship analyzing how algorithmic systems perpetuate systemic biases against disadvantaged demographics and compound historic injustices. A core concern is that algorithmic systems reproduce and amplify existing dynamics of exclusion, given their reliance on learning statistical patterns from society. Mitchell et al. [201] argue that as algorithms are designed to match the status quo, the status quo that disadvantages minorities gets replicated into automated decisions. For instance, predictive policing tools trained on datasets of overpoliced neighbourhoods amplify those patterns by sending more officers back based on biased data. Similarly, Feinstein [205] finds that algorithms encoding gendered labour segregation trends intensify occupational gender divides by steering women towards lower-paid recommended jobs. It pressures disadvantaged groups into conformity with marginalized roles rather than providing opportunities for advancement. In addition, algorithms obscure the roots of social problems in structural forces by framing issues abstractly through data correlations. Benjamin [186] states that AI-driven interventions around problems like poverty or inadequate healthcare hardly situate political and economic issues, overlooking the responsibilities of institutions and power. The "technosolutionist" worldviews that position AI as a problem-solver to complex social issues disregard the downsides of technology. Green [206] cautions that unless combined with genuine political will to undo the harms of historical injustice, simply introducing new tech systems into broken institutions only amplifies damage to disadvantaged groups.

Many critics argue that the core need is empowering marginalized communities to shape technology governance, not just technical systems. Failure to implement democratic oversight enables tools deployed in an unjust status quo to worsen outcomes. While AI could assist social progress, achieving tangible equity requires indicating that existing social orders enabling widespread biases are unacceptable. Merely optimizing technology to operate smoothly within them excludes crucial reforms. Further, the uncritical application of algorithms risks digitally reproducing social divides. Even AI aimed at fairness goals often formalizes differences into explicit data categories, reviving socially constructed divisions like race deemed illegitimate bases for decisions [186]. Similarly, personalized machine learning can segregate users into filter bubbles that consolidate their beliefs and conditions [207]. Such fragmentation risks impeding the solidarity needed to further equality. Overall, ensuring AI promotes social justice requires thoughtfully assessing how automated systems could entrench barriers between groups beyond just targeting explicit bias. Well-meaning technical fixes that formalize disadvantage may paradoxically undermine social integration and mobility necessary for just societies.

3.1.2.3   Lack of Transparency and Accountability

In addition to direct discrimination, algorithmic bias creates harm through an absence of transparency regarding how systems function and make decisions. The opacity of many modern AI techniques prevents accountability for potential unfairness or errors, violating norms of due process and openness vital for democratic governance over technology affecting society. This section will analyse biased algorithms' transparency and accountability deficits, arguing that their opacity undermines recourse and obstructs scrutiny needed to promote justice. In many contexts, like predictive policing, social media feeds, and credit lending, the algorithmic systems impacting people's lives are proprietary corporate secrets. Companies tend to block access to models and data, claiming intellectual property rights and competition risks of disclosure. This intentional secrecy means those harmed by algorithmic decisions have no recourse to understand outcomes affecting them. It also prevents external auditing for bias, as researchers must reverse engineer systems with limited access. Additionally, the technical complexity of many AI systems creates an illusion of objectivity that hides potential bias. Machine learning models appear as impenetrable black boxes generating predictions mechanically from data, leading practitioners and the public to view algorithms as impartial and neutral arbiters [43]. However, this masks how models conceal layers of problematic assumptions and choices made by developers that often result in biased behaviour. The view of algorithms as neutral "math" removes responsibility for scrutinizing their fairness, allowing harms to remain unchecked behind a veneer of objectivity. Many AI systems' technical properties also prevent explaining specific decisions made algorithmically. With techniques like deep neural networks, even developers struggle to articulate the model's reasoning for a given outcome [198]. Such opacity becomes highly concerning when AI informs impactful decisions like prison sentences or credit lending. Subjects have no ability to examine the reasoning that shaped pivotal life choices. Further, the dynamic complexity of algorithms poses challenges for regulators seeking oversight and auditing to tackle bias issues. Researchers emphasize that AI requires new forms of accountability different from past software, as algorithms constantly update based on new data, lacking static code to inspect [175]. Continually evolving models demand ongoing audits, but companies evade transparency. Governments must compel greater openness from AI creators, enabling review by outsiders for the greater good of the public. However, companies resist disclosing lucrative models, and experts struggle to interpret opaque systems. Meaningful accountability requires creative technical and policy solutions, balancing innovation with public scrutiny. At present, AI too often represents black boxes, distorting society behind veils of objectivity.

3.1.3 Towards Fairer and More Accountable Algorithms

3.1.3.1 Techniques for Bias Mitigation in Data and Models

Achieving fairness and accountability in artificial intelligence (AI) requires concerted efforts to detect and mitigate biases throughout the machine learning pipeline. Researchers have developed various techniques spanning data collection, data processing, model training, and post-hoc techniques to reduce algorithmic biases related to attributes like race, gender, and age. Here, we focus on analysing leading bias mitigation approaches and discussing their mechanics, limitations, and value in making AI systems more socially just. Representational bias in datasets used for training models is another source of algorithmic bias. One approach is balancing different groups' representation during data collection to prevent skews. It may involve targeted oversampling of underrepresented groups and classes so the model has sufficient diverse training examples [184]. However, such balancing must ensure that all examples and groups reflect the true base rates and distributions, or it could introduce reverse biases. In assessing model fairness, optimizing overall accuracy risks overlooking disparities in performance across different subpopulations. Checking for imbalanced errors or distortions separately for each impacted group is thus critical and may flag otherwise hidden biases [208]. However, defining coherent groups raises challenges of categorizations, which require ongoing consultation with affected individuals and communities. Going beyond internal testing, participatory algorithm audits by citizens and public interest groups also offer accountability benefits [197]. Allowing societally diverse external evaluators to probe algorithms can surface biases and harms that are not visible to developers and bring community values into governance.

Overall, while promising, technical interventions can only partly address bias rooted in wider social forces and history. Techniques like data balancing may improve outputs but leave unjust data ecosystems intact. Lasting solutions require complementary policy changes and democratizing development [209]. Combining technical bias mitigation with inclusive teams, community participation in design, and public transparency remains imperative for ethical AI. Finally, mitigating unfairness in algorithms will necessitate deploying both technical solutions such as adversarial learning, alongside social-political reforms that challenge entrenched power imbalances that perpetuate historic marginalization. However, concerted, multidimensional efforts hold promise for developing AI that upholds principles of justice and avoids reinforcing inequities. The key will be sustaining commitments to fairness even when they require compromising on metrics of efficiency and performance for the greater social good.

3.1.3.2 Diversifying Design Teams

One crucial intervention needed to mitigate bias and promote fairness in artificial intelligence (AI) systems is diversifying the design teams involved in developing and deploying algorithms. As discussed before, the current lack of diversity among AI researchers and engineers contributes significantly to problems of algorithmic bias and discrimination. Introducing diverse perspectives into development can surface problematic assumptions, prevent blindspots, and lead to more inclusive, ethical algorithms. Those building and governing AI systems represent a narrow demographic slice, with little representation from marginalized communities most impacted by algorithmic decisions. In the US, only 14% of AI researchers are women, while African Americans and Latinos each make up only 2–4% of the field [210]. This homogeneity feeds directly into biases and harms. Developers' limited perspectives make it hard to notice exclusions or question shared assumptions [208]. For instance, predominantly male medical AI teams building biased diagnostic tools or privileged engineers overlooking social contexts shaping minority experiences. However, evidence shows that diversity significantly improves algorithmic fairness. In one study, computer science teams with female members created algorithms with 40% lower gender bias compared to all-male teams [211]. Similarly, LGBTQ+ representation in developer teams reduces biases against queer communities being encoded into systems. Diversity prompts consideration of a wider range of user experiences and needs.

How can diversity in AI design teams be expanded? Given extensive education and career barriers faced by underrepresented groups in technology fields, interventions must occur across the pipeline. Providing girls, minorities, and low-income students more access to STEM education is a long-term foundation. Understanding the value of participatory design is crucial - directly including impacted group members as partners in the development process. Gathering input from marginalized communities missing in the formal workforce can provide indispensable perspectives on potential harms. This more immediate representation complements expanding formal diversity. There are challenges to diversifying AI, including cultural resistance within the tech industry. Some argue that considering social implications beyond accuracy could undermine technical progress. However, this frames diversity and ethics as opposing technical excellence, which evidence dispels [193]. Holistic technical-social approaches are essential. There are also risks of pseudo-inclusion that fail to transfer meaningful decision-making power.

3.1.3.3 Community Participation and Oversight

In addition to technical interventions, cultivating community participation and public oversight of artificial intelligence (AI) development and usage provides vital accountability towards combating algorithmic bias. Directly engaging impacted groups in the design process and enabling external monitoring of AI sys-

tems by civil society institutions are crucial complements to technical bias mitigation [209]. This section argues for the importance of community inclusion and
oversight, analyzing their role in making algorithms fairer and more democrati-
cally governed. Seeking input from members of groups affected by algorithmic
decisions as active partners in the design process can bring invaluable perspectives exposing potential harms to the surface. Technologists often miss key social
contexts that participants from impacted communities might effortlessly identify
from lived experience. For instance, minority software testers Critique and Code
have uncovered significant racial biases in healthcare apps and facial recognition AI missed by developers. Deeply engaging users promotes accountability.
Participatory design also skews innovation towards human centricity, not just
raw technical performance. Directly including marginalized individuals ensures
that technologies reflect collective needs, not just elite pursuits. It signals that
domains like justice and welfare should answer to democratic values, not proprietary interests [206]. Without community participation, even well-intentioned
technical fixes risk overlooking social realities. However, there are challenges to
participatory design as meaningful participation also requires transferring design
authority, not just symbolic consultation. Companies may limit engagement to
avoid hard trade-offs between profit-making and ethics. Sustained oversight requires funding and expertise development among public interest entities as well.
Community inclusion in shaping AI development paired with independent auditing of algorithmic systems offers crucial pathways to mitigate bias and promote
justice in technology. Purely technical interventions cannot address complex so-
ciocultural roots of discrimination encoded into AI. Deep and sustained engagement with impacted groups provides vital missing perspectives. Combined with
enabling civil society oversight, these participatory approaches can ground algorithmic accountability in democratic values and human rights.

3.1.3.4 Regulations and Public Policy Interventions

In addition to technical solutions, achieving more ethical artificial intelligence
(AI) requires regulatory and public policy interventions to mandate fairness, accountability and transparency. Self-regulation by the technology industry has
proven insufficient, necessitating legal reforms and government action to protect public welfare and human rights against unethical harms from algorithms
[175]. Targeted regulations can provide crucial oversight and steer innovation
in socially responsible directions. One approach is updating anti-discrimination
laws like the Civil Rights Act and Equal Credit Opportunity Act to encompass algorithmic decisions. As algorithms increasingly replace human decisions
across sectors like employment, housing, and lending, our legal frameworks must
evolve to outlaw unfair bias and discrimination regardless of its form. They provide clear statutory grounds to contest biased algorithms as rights violations
would empower oversight. Governments can also institute pre-deployment test-

ing and auditing mandates for public and commercial AI systems that pose significant risks of harm to vulnerable populations through potential bias. Establishing clear legal liability for organizations that deploy algorithms causing demonstrable harm through unfair, biased outcomes provides accountability. The threat of fines, lawsuits, and remedies for provable damages incentivise businesses to scrutinize systems thoroughly for bias before wide usage. However, care must be taken to balance liability while fostering innovation. Combined with enhanced public transparency into models, liability for algorithmic harms constructs vital social guardrails aligning innovation with ethics. While voluntary ethics are important, targeted regulation and public policy are indispensable to ensure algorithmic systems respect principles of justice, non-discrimination and accountability. With thoughtful design, interventions can balance public welfare with technological progress—however, unrestrained AI risks intensifying historic inequalities. Oversight mechanisms instituting transparency and consequences for harm are imperative.

3.1.3.5 Importance of Sustained Ethical Commitment

Achieving truly fair and accountable artificial intelligence (AI) requires more than technical interventions or one-time auditing. It demands an ongoing, sustained commitment to centring ethics and justice at all stages of technology development and governance [212]. The absence of sincere dedication to values, even when they conflict with goals like profit or efficiency, bias, and discrimination, risks recycling endlessly within AI systems. From problem formulation to data collection, model building, evaluation and deployment, ethical considerations must shape each step of engineering workflow. Rather than treat ethics as an afterthought or public relations matter, values like fairness, accountability and transparency must be foundational constraints on design choices. Such actions demand sincere self-reflection on assumptions and examining how every decision impacts vulnerable communities. Operationalising ethics requires embedding it into organizational culture. Truly committing to algorithmic justice also often requires deliberately trading off some technical accuracy or performance to adhere to moral principles [193]. For instance, while removing sensitive attributes from data or balancing representation across groups can reduce predictive power, a technical compromise is achieved rather than only optimization. An ethical commitment to AI also cannot end after models get deployed. Biases and harms often emerge only in real-world use. To address these, sustaining accountability to users via ongoing monitoring, participatory auditing, and transparency even after deployment is essential [197]. Accountability must persist throughout the system lifecycle and not be viewed as a pre-deployment checklist item to sustain a feedback loop and improve alignment with social good over time. Given that AI risks exacerbating social inequities, what matters most is the persistent, holistic commitment to upholding ethical values like justice, even when inconvenient. This requires sincerely integrating ethics into organizational cultures and workflows, making trade-offs to moral ends, and sustaining accountability post-deployment. Fairness is not achieved through technical interventions alone without underlying moral conviction orienting their application. Sustained ethical commitment remains imperative.

3.1.4 Role of Ethics and Values like Justice and Non-discrimination

3.1.4.1 Ethical Foundations for Algorithmic Fairness

As algorithms and predictive analytics become entrenched across finance, healthcare, criminal justice and other high-impact sectors, there is escalating concern that machine learning models could perpetuate injustice and violate civil rights [213]. Research into "algorithmic fairness" seeks mathematical formulations to build ethical principles like fairness directly into the design and training of automated systems. However, statistics alone cannot capture complex sociomoral concepts. Developing robust ethical foundations for algorithms requires introducing core values like fairness and non-discrimination, analyzing associated moral philosophies that lend legitimacy to these principles, and building convincing arguments for embedding ethical reasoning within applied data science work. This section will provide a rationale for insisting that algorithm designers wrestle with fundamental questions of justice before attempting to optimize purely technical metrics. Algorithmic fairness literature frequently concentrates on three preeminent ethical values intended to spur the creation of more just and socially beneficial algorithmic systems: The first core ethical principle centres on justice - assigning resources, opportunities and decision-making power in accordance with shared conceptions of fairness mediated through public institutions. Algorithmic justice thus implies that automated systems should yield outcomes widely perceived as morally right, legitimate and equitably impacting all groups across diverse societies. Closely linked to justice, fairness constitutes a complex evaluative standard that concentrates broadly on eliminating unjustified prejudice to promote equitable treatment. Algorithmic fairness explores technical methods for removing unfair statistical biases to build systems that avoid marginalizing vulnerable groups or denying individuals opportunities based on ethically irrelevant attributes like race, gender or disability status.

These core values disseminated within algorithmic fairness discourse connect directly to major schools of Western moral philosophy and ethical reasoning that construct sophisticated arguments regarding why principles of justice, rights and equality hold intrinsic importance for creating better societies. The utilitarian approach of maximizing overall well-being and aggregate happiness across the full populace supports designing algorithmic systems to benefit most people. Jeremy Bentham, an early theorist, promoted how just governments and policies require impartial calculus measuring utility impacts upon all citizens, a stance later economists and computer scientists have applied to ensure algorithms increase societal welfare rather than accrual of power or profit to narrow interests [200]. Social contract tradition concentrates on principles of justice emerging from rational consensus through public deliberation among free individuals. John Rawls [214] famously argues that fairness results from negotiations behind a "veil of ignorance" about one's status in society. Similar impartial reasoning on algorithmic impacts, especially for disadvantaged groups, accords with these origins of justice as rightful treatment determined via inclusive moral dialogue and debate. While technical studies on mitigating algorithmic biases and architectural strategies for enhancing system transparency abound, ethics remain alarmingly sidelined in much computer science research and application [215]. However, establishing conceptual clarity and persuasive reasoning around core values at stake is critical before quantifying notions like fairness or encoding them into computable models. Only through analyzing foundational moral philosophies supporting principles against domination, marginalization and discrimination can data scientists build legitimacy for algorithmic systems and foster public trust that technical interventions will truly remedy ethical concerns rather than privilege powerful stakeholders. Statistical models cannot intrinsically determine the ethical weighting of different harms, privacy infringements, or discrimination types without philosophical principles applied via external moral judgment. Debates over algorithmic biases reflect disagreements in values, not just math, requiring constructive dialogues between technical and ethical experts to reach contextualized conclusions [212].

Justice, fairness and non-discrimination constitute bedrock moral convictions for contemporary democracies demanded of emerging algorithmic systems. Through constructive collaboration, data scientists cognizant of moral foundations and ethicists literate in computational tools can jointly cultivate innovative, ethical techniques that embed time-tested values into algorithms designed to empower rather than subjugate humanity.

3.1.4.2 Justice as Fairness in Algorithmic Systems

As algorithmic systems powered by machine learning are deployed across critical social domains like criminal justice, employment, healthcare, and finance, concerns mount over discriminatory impacts and unfair outcomes disproportionately affecting disadvantaged groups. In response, computer scientists explore technical interventions to remove biases and fulfil vague ethical objectives around fairness, accountability and transparency. However, lacking rigorous philosophical grounding, these efforts risk failure or misdirection. The political philosopher John Rawls provides essential support through his seminal work developing a "justice as fairness" theory deduced through rational principles widely applicable to algorithmic systems. In A Theory of Justice, John Rawls [214] puts forth an ingenious thought experiment asking what principles of justice free and rational people would agree to under fair conditions for social cooperation where no one knows their eventual status, talents or position in society. Under this "veil of ignorance", self-interested bargaining cannot influence the results towards unfair accumulation of power or wealth. The resulting two principles of justice demand:

- Equal basic liberties and rights for all citizens.

- Fair equality of opportunity and the "difference principle."

This system powerfully requires algorithmic systems to uphold civil rights, avoid bias that restricts opportunities for the marginalized, and regulate inevitable inequalities to benefit vulnerable groups maximally. Beyond formal rights, Rawls argues justice entitles all citizens to equality of opportunity regardless of social or economic status. algorithmic tools utilized in areas like hiring, admissions, and lending must face scrutiny accordingly if they narrow prospects for marginalized groups or negatively interact with existing background inequality sources through no fault of applicants. Impartial fairness fundamentals evaluate the ethicality of algorithmic systems based on experienced injustice from affected groups, not aggregate efficiency or productivity. Risk assessment tools, predictive models and other algorithms requiring such consequentialist trade-offs should thus undergo review processes akin to Rawls' original position thought experiment behind a veil of ignorance concerning which groups bear the burdens.

3.1.4.3   Algorithmic Discrimination as Injustice

As algorithms and predictive models inform high-stakes decisions across finance, employment, criminal justice and more, evidence mounts that seemingly neutral automated systems can produce unfair, discriminatory and unjust outcomes through unexamined biases [203]. However, technical literature frequently lacks coherent definitions and rigorous ethical reasoning on different bias types and their implications within applied fields. Synthesizing conceptual and empirical insights proves critical. This section will define key forms of algorithmic discrimination grounded in justice theory and examine sources of unfairness violating core principles of impartiality and proportionality. Establishing algorithmic discrimination as manifest injustice lays essential groundwork for interventions to align automated decisions with ethical standards of fairness and accountability. While often used imprecisely within technical discourse, the following hierarchical taxonomy classifies major algorithmic bias varieties:

- Preexisting biases - Historical discriminatory patterns and structural inequality manifesting in training data that algorithms learn and perpetuate [216].

- Technical biases - Model overreliance on spuriously correlated features leading to unfair generalizations and inertia [184].

■ Emergent biases - Unanticipated discriminatory effects arising from context shifts between training and implementation [217].

Coupled with appropriate methodology, this schema helps trace biased algorithms back to sources of injustice embedded within data. At the core, biased, discriminatory algorithms contradict the primary attributes of just decisions aligned with a fair and inclusive society. Three principles include:

■ Impartiality - Decisions influenced by ethically irrelevant attributes like race or gender contradict impartiality vital for justice.

■ Proportionality - Algorithmic systems that inaccurately categorize or unevenly restrict opportunities for minority groups challenge the proportionality components of just treatment.

■ Progressive Equality - Biased algorithms that reinforce historical oppression against disadvantaged communities violate equitable standards of justice.

Through these failures of impartiality, proportionality and equality, algorithmic discrimination constitutes manifest injustice.

Several high-profile cases have revealed how biased algorithms unleash devastation on vulnerable communities already suffering marginalization and discrimination:

■ Predictim - A private firm offering AI life coaching advice was found illegally using race as an input variable to make broad generalizations predicting children's criminal tendencies later in life with no scientific grounding [192].

■ Healthcare - An algorithm meant to identify patients needing extra healthcare interventions was found twice as likely to flag black patients due to confounding chronic illness proxies skewing predictions along racial lines without cause [203].

■ Hiring - Amazon scrapped an algorithmic resume screening tool shown to discriminate against women by penalizing graduates from all-women colleges [204].

In contexts from criminal risk assessments to medical diagnoses and hiring, algorithmic discrimination contravenes basic standards of justice through biased systems that violate rights and reinforce historical oppression against already vulnerable populations. By framing the problem through rigorous definitional analysis and moral philosophical critique, pathways emerge to challenge algorithms on the grounds of injustice and compel reforms that promote impartiality, proportionality and equality consonant with the core value of fairness.

3.1.4.4   Cultivating Ethical Algorithms

Algorithms and automated decision systems rapidly proliferate across high-stakes social domains, fueling rising apprehension over unfair biases, discrimination, and unjust outcomes. In response, academics and activists urge significant reforms throughout the algorithmic lifecycle to embed ethical values like justice and non-discrimination directly into emerging sociotechnical systems. Realizing this vision demands exploring multiple complementary pathways for developing ethical algorithms. This section surveys promising strategies, including participatory design, diverse development teams, transparency regulations, and needed policy interventions. Synthesizing these solutions points towards a multifaceted agenda centred on stakeholder empowerment and continual critical engagement to build just algorithms that earn public trust and enable human flourishing. Expanding who builds and governs sociotechnical systems proves vital for upholding pluralistic values. Community-based participatory design methods empower those affected by algorithms and structural interventions like external algorithmic auditing boards with representation from marginalized groups similarly work to make algorithms more responsive through ongoing feedback channels [197]. Additionally, diversity among internal engineering teams allows different perspectives to highlight potential sources of unfairness or discrimination overlooked by homogeneous tech workers. However, lasting change requires meaningful incorporation of impacted groups and advocates into corporate or public sector decision-making, not just ad hoc consultation [186]. Enforcing transparency remains crucial so external auditors and the public can critically evaluate algorithmic systems for biases and unfairness. To avoid confusing the public with technicalities, regulatory reforms should mandate that companies detail their entire development pipeline and fund independent third-party auditing programs [218].

To truly undo the oppressive structures that allow algorithms to marginalize disadvantaged groups further requires policy interventions. Researchers have proposed various solutions, including empowering public interest lawsuits, instituting affirmative consent or fairness certification regimes before deploying algorithms, and creating specialized regulatory agencies to continually assess and mitigate emerging concerns [219]. Cultivating fundamentally ethical and just algorithms requires multidimensional efforts bridging technical, social and regulatory spheres. Public participation, diverse development teams and transparency policies should constitute the default foundations of algorithmic infrastructure. With legal backing and public sector oversight, possibilities emerge to make algorithms serve justice rather than undermine it.

3.1.5 Outlook for Future Research and Policy Directions

Algorithms and predictive models now mediate countless impactful decisions in people's lives, from criminal risk assessments to financial lending to diagnosing medical conditions. However, ample evidence shows that seemingly neutral algorithmic systems can produce unfair, discriminatory and unjust outcomes due to unexamined biases and exclusions embedded within historical training data or insensitive system design choices [184]. Addressing this modern problem constitutes both a technical challenge and a governance crisis demanding responses that balance innovation with appropriate safeguards guided by ethical reasoning on emerging challenges still undergoing dynamic understanding. This section will examine promising horizons spanning research insights, participatory interventions and policy reforms needed to cultivate accountable algorithmic infrastructure aligned with priorities of justice and non-discrimination against vulnerable populations. Technical literature on biased algorithms remains in an early yet highly active exploratory phase, with extensive opportunities to advance practical progress on inherent challenges. Three research pathways warrant highlighting:

■ Inclusive Assessment - Novel auditing methods like algorithmic impact assessments, benchmark suites and stress testing approaches developed collaboratively with impacted groups can expose unfair model behaviours missed by standard evaluation [197].

■ Responsible Innovation - Interdisciplinary studies on context factors influencing real-world effects when deployed, participatory design procedures integrating stakeholders, and policy foresight studies using simulations promise responsible innovation ecosystems for developing and testing algorithms [220].

■ Holistic Bias Mitigation - Rather than seeking narrow technical fixes, researchers urge holistic improvement across the entire machine-learning pipeline, from curating inclusive datasets to causality-aware modelling techniques to enable robust and ethical algorithms [208].

While empirical insights accumulate, policymakers must formulate governance regimes, ensuring companies and institutions deploy algorithms responsibly. Multiple complementary policy directions hold promise:

■ Risk and Impact Reviews - Drawing lessons from environmental registries, agencies can mandate filing notices detailing the context of use and risk models before deploying algorithms that substantially impact public welfare [219].

■ Standards - Voluntary consensus standards or required certifications on algorithmic accountability with input across industry, academia and civil society can entrench best practices [197].

■ Explainability Laws - Enhanced transparency regulations forcing companies to detail complete model parameters, data compositions and bench-
marked group performances support external audits and contestations of
unfair systems [218].

■ Oversight Institutions - Public sector bodies providing ongoing governance and investigative authority over algorithms utilized in regulated
sectors like banking or healthcare steward just development across contexts lacking market pressures [221]. Through these multifaceted policy
reforms grounded in emerging interdisciplinary insights on responsible
innovation, the outlook improves for constructive societal guidance of
transformative algorithmic systems towards equitable and ethical outcomes. The challenges biased algorithms present resonate deeply with
enduring questions of justice and fairness within liberal democracies.
While easier to ignore during initial unchecked deployment, ramifications from unfair algorithmic systems that reinforce historical oppression become evident across numerous fields. Lasting progress relies on
research that engages impacted groups, surfaces unintended harms, and
advances solutions to balance complex trade-offs. Simultaneously, constructing policy regimes and governance institutions prepared to steward
continued algorithmic advances through evidence-based standard setting
and review processes offers a pathway for realizing societal benefit rather
than merely accruing power or profit. With ethical reasoning, inclusive
scientific exploration and accountable oversight now coming into focus,
tentative optimism emerges on harnessing algorithms' immense capabilities to empower communities previously denied opportunities to thrive
meaningfully.

3.2 Privacy and Surveillance: Balancing Individual Rights and Benefits of AI

3.2.1 Understanding Privacy Harms from AI Surveillance

3.2.1.1 Privacy and the Evolution of Concepts like Social Privacy

Rapid advances in big data and artificial intelligence significantly empower both
public and private surveillance capacities, threatening core civil liberties protections. However, privacy constitutes a complex concept encompassing many
distinct forms and interests evolving through legal, technological and social contexts [222]. This portion explores how definitions and perceptions around privacy adapt to emerging surveillance infrastructures, analyzes novel threats like
social privacy leaks, and lays the groundwork for discussing appropriate countermeasures in subsequent sections. Legal and philosophical debates have developed overlapping characterizations of privacy centred on individual control,

limited access, and interference prevention. However, new technologies with extended monitoring capabilities increasingly challenge traditional spatial, informational, and social boundaries of privacy. The modern interpretations acknowledge privacy not as an isolated good but as an essential foundation for identity formation, social participation, and civic discourse [223]. Complex algorithmic readings about people from stealthily collected data discourage intuitive understandings and demand a broader understanding of relational dynamic privacy elements [224]. A particularly important concept in this discussion is social privacy - the ability of groups to collectively associate, speak freely, and develop secure subcultures under limited scrutiny from authorities or corporate interests [225]. Social media increases public visibility and tracks informal connections, eroding social privacy protections and enabling mass opinion analysis or behaviour. However, regulation and technical interventions primarily concentrate on individual data rights, neglecting collective impacts from information ecosystems profiling to shape how groups form opinions, make decisions, or become targets of repression [39]. Reassessing and enhancing social privacy thus constitutes a critical frontier in mitigating privacy risks from algorithmic systems. Among many documented impacts, three harms helpful in conceptualizing the rights-based critiques

- Discrimination – predictive analytics, facial analysis, and behavioural tracking systems discriminatorily target disadvantaged groups by including biases, proxies, or historic prejudice embedded in model training data or institutions [221].

- Manipulation – profiling individuals through captured online activities, friendship networks, and location history to micro-target information in a calculated manner to benefit commercial or political interests.

- Information asymmetry – the knowledge gap between what companies and state agencies assume about citizens through extensive monitoring compared to what users can reasonably determine about such entities' surveillance, analysis, and deployment disempowers individuals subjected to one-sided transparency [226].

In the face of ubiquitous algorithmic surveillance systems, society requires expanding conceptions of privacy that encompass social group dynamics and central concerns around discrimination, manipulation, and information asymmetries. Establishing these foundations sets the stage for critically evaluating contextual uses of privacy-violating.

3.2.1.2 Case Studies of Privacy Controversies (e.g., Clearview AI, Cambridge Analytica)

Beyond abstract definitions or regulatory frameworks, real-world privacy controversies sparked by emerging technologies provide instructive details into actually occurring harms. Analyzing notable examples like covert face recognition outfit Clearview AI and the Facebook—Cambridge Analytica political micro-targeting scandal reveals tangible damages to individual rights and democratic institutions from unchecked commercialization of invasive data systems. This practical grounding informs policy responses and organizational reforms towards restraining data extractivism and constructing accountability safeguards at scale. Clearview AI constitutes a mass facial recognition startup scraping over 3 billion images from public web sources and social media sites without user permission. It then enables instant searches matching photos to identity profiles and contact information for law enforcement and private sector clients through an automated database architecture [227]. This entirely covert endeavour violates assumed privacy boundaries around biometric data collection, supporting functionality deeply concerning civil liberties advocates. Known externalities include misidentification and over-policing harms to marginalized groups and normalising ubiquitous public tracking, eroding freedom of association [228]. Though currently facing lawsuits, Clearview epitomizes risks from security-oriented AI surveillance developed and deployed unilaterally by private companies' absent ethical review.

Alternatively, Cambridge Analytica (CA) exemplified privacy damages from targeted computational propaganda platforms explicitly designed to shape voter decisions. By acquiring millions of Facebook users' personal data without consent and combining this with purchased consumer datasets, CA constructed psychographic profiles to microtarget information calculated to manipulate beliefs and behaviours actively [229]. This privacy violation for mass persuasion goals reveals risks of online behavioural tracking given inadequately secure commercial data ecosystems, lax oversight, and gaps between data rights regimes spanning continents. The CA scandal spotlighted needed reforms around social media privacy settings, platform accountability, international data flows and modernizing legal consent standards for data sharing in the public interest rather than solely for corporate profit. From Clearview's insatiable scraping of identification markers to Cambridge Analytica's deception and targeting, real cases of privacy controversies highlight tangible ways emerging sociotechnical systems fail to protect individuals' civil liberties, autonomy, dignity and trust. Understanding privacy loss and exploitation dynamics must inform solutions balancing innovation with individual rights.

3.2.2 Arguments for and Against AI Surveillance

3.2.2.1 Benefits: National Security, Law Enforcement, Personalized Services

While privacy, autonomy and consent concerns characterize much of the discourse on emerging AI surveillance systems, advocates argue that such technologies generate benefits for public safety, national security, and service customisation that outweigh hypothetical risks of misuse or overreach. This section examines the leading arguments commending social advantages from increasingly data-driven algorithmic monitoring, profiling and predictive analytics across law enforcement investigations, border security, and consumer product recommendations. However, these professed benefits prove largely unfounded given disproportionate impacts on marginalised communities, institutional opacity that frustrate accountability, and alternative policy choices supporting innovation with civil rights protections. Government officials advocate harnessing AI surveillance technologies like facial recognition, predictive policing analytics, backdoor encryption access, and expansive digital communications monitoring to further national security and law enforcement objectives [230]. Beyond efficiency gains, such systems supposedly enhance investigation success rates, reduce ethically questionable profiling practices, and proactively identify potential threats through pattern analyses similar to human analysts but infinitely more scalable. However, most claims of indispensable security benefits lack empirical backing or contextual validity. High-profile cases of counterterrorism algorithms misclassifying ethnic minorities or peaceful protestors further contest the accuracy claims [224]. Consumer contexts like shopping, entertainment or social media also increasingly utilize AI surveillance to discretely profile user behaviours, connections, and preferences across platforms. Such customised services driven by data extraction improve user experiences by anticipating desires, matching tastes to products, and adapting interfaces to individual needs. However, this omits the manipulative and adversarial optimization governing commercial algorithmic systems, which are increasingly incentivised to maximise engagement metrics benefitting the platform over user wellbeing [231]. Users face constrained choices and dark patterns, influencing behaviours against intentions rather than robust user-centric personalization.

While promises of security and customization help drive unchecked adoption of pervasive monitoring technologies, investigating actual system logic and disproportionate impacts reveals that AI surveillance essentially serves institutional power rather than citizens or consumers.

3.2.2.2 Critiques: Infringements of Civil Liberties, Chilling Effects, Consent Issues

Countering claims around security and efficiency gains, critics urgently highlight numerous downsides to unrestrained AI surveillance systems engaging in mass data extraction and retention. Beyond quantifiable privacy harms, societal damages include disturbing free expression, normalized violations of civil dignity, and the gradual erosion of autonomy rights through opaque algorithmic governance [232]. This section details three central categories of opposition to deploying pervasive monitoring programs: infringements of core civil liberties, chilling effects constraining pluralism, and consent insufficiencies undermining consumer or citizen interests to corporate and state actors acting unilaterally without meaningful public input or individual control. Despite constitutional and legal privacy protections in democracies, many AI surveillance systems categorically infringe on liberties ranging from privacy and anonymity in public spaces to associational freedoms and privileges against self-incrimination [233], particularly concerning the potential violations of due process owing to the reliance on statistical correlates, risk scores, racial identifiers, or error-prone computational categories. Similarly, law enforcement attempts to mandate or hack encryption mechanisms undermining rights protecting personal communications and thought.

Even without direct censorship, awareness of persistent monitoring and behaviour profiling have chilling effects on free expression, inquiry, and creativity [234]. When combined with automated content analysis, which is incapable of understanding contexts like humour or irony, data-driven surveillance discourages intellectual risk-taking and nonconformity. Facing heightened scrutiny or moderation, many individuals tend to self-censor views on sensitive issues to avoid platform penalties or undue profiling. Such forced conformity and disciplining directly impact idea diversity and open discourse foundational to pluralistic, functioning democracies. Further, most citizens never affirmatively consent to intrusive corporate or governmental surveillance or negatively respond to an opportunity to understand the scope of the surveillance. Opaque algorithms utilise data scraped, purchased, or hacked without approval, thus cheating the unaware user permissions [215]. Moreover, despite publicity catastrophes like Cambridge Analytica showing platforms cannot safeguard personal data, consumers retain minimal agency to protect information because the concentrated industry structures and internet architectures permit seamless behind-the-scenes data sharing. From core civil rights to participatory self-determination, AI surveillance systems arguably inflict more societal harms than asserted benefits in the absence of robust public oversight and consent controls. Constructive reforms must address this blatantly disproportionate power dynamic central to surveillance.

3.2.3   *Towards Solutions Balancing Privacy and Innovation*

3.2.3.1   *Policy Interventions: Data Protection Laws (GDPR), Algorithmic Transparency*

While AI surveillance advances through global technology firms and data broker ecosystems, policy responses lag behind, impacting civil rights and autonomy. However, promising interventions, including strengthening data protection statutes and mandating supply chain algorithmic transparency, offer pathways to restore balance against unchecked corporate and governmental monitoring systems. This section focuses on leading policy proposals around comprehensive data rights guarantees and transparency requirements compelling disclosure of system designs and application programming interfaces (APIs) to enable external auditing. Though complex to implement, such multifaceted reforms incentivise accountable development processes serving user privacy over unilateral data exploitation. Since architectures of digital technologies align towards perpetual data extraction in the absence of economic or legal constraints, one vital policy direction involves enhancing data protection guarantees through expanded privacy laws, giving individuals control over behavioural, transactional, and demographic information involuntarily collected about them by companies or institutions [235]. The European Union's General Data Protection Regulation (GDPR) is the pioneering attempt at ensuring access, review, deletion, and transparent consent requirements for data handling paired with significant fines for violations. Furthermore, the state of California recently instituted the California Consumer Privacy Act (CCPA), similarly compelling disclosures and opt-out abilities covering technology firms. Though insubstantial, such policy innovations pressure improved corporate accountability. Additionally, directly targeting the growing role of algorithms in managing or utilising scraped personal data proves essential for meaningful oversight. Researchers advocate amended transparency laws forcing companies to fully detail proprietary algorithms and models and provide debuggable access through open application programming interfaces (APIs) to enable external audits [197]. Constructive transparency further requires disclosing complete development pipelines, documenting data sources and compromise decisions, detailing performance benchmarks across user segments, and funding independent auditing programs. Through strengthening legal data rights and requiring state-of-the-art user protections to become embedded in systems by design rather than as an afterthought, policymakers can overcome lagging technical accountability and minimise the harms from AI surveillance systems' uncontrolled expansion across society.

3.2.3.2 Responsible Innovation - Privacy by Design, Ethics Review Boards, Codes of Conduct

Cultivating accountability around AI systems necessitates moving beyond external policy interventions towards directly embedding ethical practices and priorities into the design lifecycles of algorithmic products and services. For responsible or ethical innovation, it is essential to approach privacy and values through design, establishing ethics governance boards, and creating codes of conduct compelling to consider moral duties above the single-minded pursuit of technical capability [236]. This section analyses organisational reforms that promise renewed user trust through institutionalising impact assessments, oversight procedures and value alignment practices to steer technology firms away from single-minded data exploitation pathways. Processes that tackle ethical considerations early during goal formulation, use case selection, dataset curation, and prototype testing stages allow holistically minimising harmful acts like privacy violations rather than attempting to append privacy as an afterthought or policy requirement. Inspired by "security by design", guidelines for privacy by design stress data minimisation, anonymity, decentralised storage and consent requirements get directly encoded into system specifications from initial design phases rather than as retrospective remedies [197]. Ethics by design practices further expand the review to address the full spectrum of social issues using participatory methods and QA testing against codified principles.

Establishing independent ethics advisory boards with sufficient technical literacy, diversity of viewpoints and rotational membership drawn from both internal employees and external advocates directly input shared accountability across organisations [212]. Charged with assessing upcoming products and key decision points against established ethical codes and issues checklists, such bodies can provide constructive feedback on social impacts otherwise discounted by pursuing narrow targets and intense competition. Further, voluntary industry self-governance through groups jointly developing codes of conduct helps embed norms against exploitative data collection or monetisation schemes that violate public expectations of proportional usage [237]. However, effective implementation relies upon independent auditors, enforcement policies penalising violations, and commonly aligning conduct codes to statutory requirements. Cultivating responsible innovation ecosystems provides encouraging possibilities for aligning incentives with user privacy through governance mechanisms internalising ethical reasoning, participatory design and accountable deliberations for product development life cycles rather than expendable expenses.

Chapter 4

Artificial Intelligence Societal Impact and Seizing Opportunities

4.1 Income Inequality - The Potential for AI to Exacerbate Economic Disparities

4.1.1 Effects of on the Job Market

Tasks that involve manual dexterity have been a primary focus for automation for quite some time now, including basic assembly line manufacturing, food preparation, and cleaning services. With the advancements in machine learning algorithms, robots are now capable of performing a wider range of sensory-motor skills that can match or even surpass humans. As a result, they can now perform more complex manual jobs, such as warehouse fulfilment, agricultural harvesting, and warehouse transportation. Technical feasibility studies indicate that over 60% of occupations, such as sewing machine operators, shoemakers, butchers, and building cleaners, are susceptible to at least partial automation [238]. The development of robotics, computer vision, and other capabilities has led to increased economic feasibility of automating manual work in the upcoming years. Certain trades, such as carpentry, plumbing, and welding, require situational adaptability, which can now be achieved through automation. Although robots complement humans in some manufacturing roles, fully automating production processes eliminates many low-skill repetitive jobs. In addition, new service robots such as Flippy, the burger-flipping bot, are now available to automate more food service roles. As technology continues to advance, the impact of automation on routine manual labour across various sectors is only expected to accelerate.

Positions in office and administrative support, such as data entry clerks, payroll processors, secretaries, and claims adjusters, are similarly susceptible to automation through the use of AI and software [48]. Machine learning has progressed to the point where it can match or even surpass human accuracy in structured information-processing tasks. Chatbots, such as Amelia, have also advanced to the point where they can understand speech and language well enough to potentially replace many customer service roles in the near future [239]. Intelligent software is automating contract management, mail sorting, and other back-office workflows, leaving only complex administrative duties that require judgment and are difficult to automate. The reduction of repetitive clerical roles resulting from automation could have a positive impact on society if it allows for more fulfilling work. However, workers displaced due to automation will require assistance transitioning to new occupations. Lower-paid administrative personnel may face more challenges in accessing such opportunities than mid-level professionals whose roles are augmented by productivity software. It is crucial to continually evaluate the impacts of automation on labour and develop policies to assist vulnerable demographics. In addition to automating routine low-skill jobs, AI is transforming many mid-level careers by partially automating work processes. Intelligent algorithms are integrated into legal research, financial analysis, medical diagnosis, and engineering design to handle data-heavy tasks while augmenting human capabilities [240]. The impact on employment in these fields is mixed.

AI enhances productivity, enabling knowledge workers to achieve more. However, this may lead to a decline in overall staffing needs. For instance, automation of routine data tasks could result in 8-10% fewer accounting and audit positions in the US, while new opportunities may arise in managing AI systems and providing strategic advisory services [48]. As expert systems take over the analysis of test results, doctors have more time to treat patients. The remaining human roles involve relationship building, communication, judgment, and oversight. It is crucial to recognize that AI transformation will significantly impact middle-tier occupations and lower-wage jobs to successfully transition the workforce.

Despite the potential for automation to displace certain tasks, it is also introducing new roles that combine AI's data-processing capabilities with human qualities such as creativity, empathy, and ethics. Engineers, designers, marketers, and other professionals are increasingly working alongside AI systems as collaborative partners [241]. User experience experts are responsible for optimizing user interaction with AI interfaces. Before actual deployment, machine learning models are supervised and validated by trainers. Meanwhile, domain specialists such as doctors and lawyers are accountable for interpreting and contextualizing

AI recommendations. The need for AI translators, educators, and ethics specialists who analyze algorithms, mitigate the risks of bias, and promote transparent societal outcomes is increasing. These roles enable a hybrid human-AI workforce model using AI as a productivity multiplier while prioritizing human needs and oversight. However, individuals without digital skills or exposure are at risk of being left behind. Therefore, intentional policies aimed at inclusive AI talent development are necessary to ensure positive future job creation.

One crucial concern regarding the impact of AI on employment is that the short-term effects of significant job losses resulting from automation may outweigh new job creation. However, historical evidence demonstrates that labour markets eventually adapt to the changes as new industries absorb the workforce. Nevertheless, research suggests that automation may initially displace up to 3% of current US jobs over the next decade, based on the observed adoption of technology [242]. The rapid retraining of all affected groups poses significant challenges. There are also concerns that the pace of AI advancement makes this era of automation distinct [243]. In the past, job losses were compensated for by the introduction of new products and services enabled by technology. Nevertheless, some economists argue that AI could disrupt new domains more rapidly than humans can adapt, emphasizing the urgency for policymakers to implement transition programs, review safety nets, and closely monitor automation impacts on workers. While smooth labour force adjustment is possible, it will require foresight and long-term planning.

The impact of automation on the job market has significant implications for income distribution. AI and technology have a significant impact on routine jobs in the middle of the skill spectrum. As a result, the labour market is experiencing polarization, with growth concentrated in high-skill analytical and technical roles and lower-skill service jobs that require situational agility [244]. High-salaried professionals witness productivity boosts while working with AI. However, low-wage food, health, and personal services rely on intrinsic human capabilities that machines lack. Unfortunately, occupations in the middle tier experience shrinkage without clear upward paths. The observed polarization of the labour market resulting from automation could potentially lead to the erosion of the middle class, exacerbating inequality. Low-wage service workers, who are not susceptible to automation, lack bargaining power for pay increases, even in tight labour markets. Policymakers are confronted with the challenge of mitigating this imbalance through the implementation of upgraded skills programs, new job creation initiatives, and potentially alternative income redistribution mechanisms in an AI economy. One of the major concerns related to large-scale job automation is the smooth transition of displaced workers to new occupations. Sudden layoffs can be emotionally and financially disruptive for individuals. Additionally, lower-income workers often have limited savings to finance retraining and job-seeking [245]. Moreover, geographic obstacles may prevent workers from accessing new industries if they do not emerge in their communities. Furthermore, individuals belonging to marginalized groups who are already excluded from existing workforces and training programs may face additional challenges in accessing new opportunities. Policies such as wage insurance, relocation assistance, and temporary basic income support could potentially alleviate the adverse effects of permanent job loss. However, governments and communities should adopt proactive economic development and training pipelines to minimize displacement shocks. Encouraging new industries to locate in areas affected by job loss could also mitigate geographic inequality. Nevertheless, achieving seamless labour force transitions to productive new work requires overcoming financial, skill, and psychological barriers on both local and societal scales.

4.1.2 Gains from AI Adoption

A robust and thriving ecosystem of start-ups has played a critical role in transforming academic research in the field of artificial intelligence (AI) into commercially feasible products and services. These small firms have demonstrated remarkable proficiency in identifying niche use cases and creating customized AI solutions for businesses. These agile start-ups have outpaced established vendors in domains such as cybersecurity, AI fraud detection, and others [246]. The availability of venture capital funding has facilitated the recruitment of scarce AI talents, even before the coveted IPO windfalls. Additionally, the acquisition of start-ups has allowed tech giants to assimilate AI capabilities, thereby expanding their offerings and market presence. The symbiotic relationship between academia and start-ups is a major driving force behind the growth and proliferation of the AI industry. Despite their remarkable ingenuity and agility, start-ups often lack the necessary resources to venture into capital-intensive sectors such as transportation, logistics, and pharmaceuticals. For instance, developing production-grade autonomous vehicle systems requires billions of dollars in investment, a feat that only industry giants like Waymo can achieve. Additionally, start-ups often encounter significant obstacles when trying to scale up globally and match the vast data access enjoyed by incumbents. While start-ups have played a crucial role in catalyzing innovations in the AI industry, their contributions alone may not suffice to achieve the desired economic diversification. Therefore, a concerted effort is necessary to ensure that the AI economy is inclusive and sustainable. The current AI landscape is predominantly dominated by a handful of large technology companies hailing from the United States and China. These behemoths, including Google, Microsoft, Amazon, Alibaba, and Baidu, possess a formidable edge over their competitors owing to their exclusive access to proprietary AI research, vast data resources, and the ability to attract top-tier engineering talent [247]. Decades of accumulated user data have enabled these companies to fine-tune their algorithms for search, recommendation, translation, and other applications. Their expansive cloud infrastructure also provides virtually unlimited computational resources for training complex AI models. Furthermore, their financial muscle allows them to recruit entire AI research labs, offering them a competitive edge over their rivals. The sheer dominance of these companies in the AI industry is a testament to their exceptional capabilities and underscores the need for a level playing field that fosters innovation and fair competition.

The formidable endowments that tech giants enjoy have enabled them to reinforce their leadership positions with each new advance in the AI industry. Even blockchain-based models, initially designed to distribute economic control across a broad spectrum of users, have seen their concentration on platforms like Amazon Web Services, thereby contributing to the perpetuation of the existing data-network effect-driven cycle. External interventions are required to break this cycle and expand access to AI resources. In the absence of deliberate policies promoting inclusivity and equitable access to AI, the existing tech hierarchy appears poised to capture a disproportionate share of the gains associated with the technology. Therefore, policymakers and industry stakeholders must work together to create a level playing field that ensures the benefits of AI are widely distributed. Online platform markets, such as those for search, social networks, and e-commerce, have witnessed the emergence of winner-take-all outcomes, where one solution dominates the majority of users. These outcomes are produced by network effects, which enable companies like Google and Amazon to improve the accuracy of their AI services. By retaining increasing numbers of users, these companies' AI services become better at predicting preferences, thereby pulling further ahead of their competitors. These self-reinforcing cycles, powered by AI-fueled personalization, result in an extreme revenue concentration in one or two providers per sector [248]. The revenue concentration in the hands of a few dominant players reinforces their position, making it increasingly difficult for new entrants to gain a foothold. Furthermore, these cycles may create a perception of an insurmountable lead, further discouraging potential competitors from entering the market. As a result, these self-reinforcing cycles may lead to a market structure that is highly concentrated, with a few dominant players controlling the majority of market share, while smaller players are relegated to niche roles.

By their nature, multi-sided networks generate dependencies that make it difficult for users and partners to switch easily. This is particularly true for large platforms such as Amazon's marketplace and cloud services, where the costs of migrating to alternative platforms are often prohibitively high. This entrenchment of dominant players in the market favours the winners of the AI innovation race, making future displacement of these players unlikely. The concentration of economic gains also manifests geographically, with tech hubs such as Silicon Valley attracting talent and investment, reinforcing the dominance of established players in the market. To address these challenges and promote competition in the market, targeted anti-trust regulations may be warranted. Such regulations may help to restore genuine competition, enabling new entrants to challenge the dominance of established players and promoting innovation in the AI industry. Developing production-grade AI systems requires significant upfront investments, which often prove to be a major stumbling block for smaller firms seeking to compete against their deep-pocketed counterparts. The development of AI systems requires vast computational power, multidisciplinary talent, and extensive data samples, which make it prohibitively expensive for most businesses [246]. Smaller players are often unable to subsidize and sustain the years of research required to trailblaze new algorithms, thereby limiting their ability to compete with established players in the market. Furthermore, paying for custom AI solutions often locks in dependencies on tech providers, thereby exacerbating the challenges faced by smaller firms. These dependencies can be complex to break, making it difficult for smaller players to transition to alternative providers, even when they become available. Consequently, the concentration of AI development in the hands of a few dominant players may create a situation where the industry is dominated by a few large corporations, thereby stifling innovation and limiting the scope for new entrants to challenge established players.

Various measures, such as government grants, shared data repositories, and cloud credits, have been proposed to democratize access to AI building blocks and enable smaller players to compete against their larger counterparts. While these measures have the potential to promote inclusivity and democratize access to AI resources, they have not been effective in denting the entrenched competitive barriers that favour dominant incumbents in the industry. The high fixed costs associated with developing top-grade AI systems appear to favour balance sheet-dominant incumbents, thereby concentrating economic gains in the hands of a few players. Consequently, the AI industry seems poised to witness the emergence of a market structure that is highly concentrated, with a few dominant players controlling the majority of market share. However, regulatory interventions can foster a less concentrated AI landscape, whereby smaller players can compete with established players on a level playing field. For instance, regulations mandating data sharing and restricting anti-competitive practices can promote competition and innovation in the industry, thereby creating a more inclusive and equitable AI landscape. Major technology companies strategically utilise AI to amplify the competitive moats guarding their business empires, further concentrating gains [249]. The vast data reserves accumulated by companies such as Google and Facebook provide a crucial input for their algorithms, which they use to improve their AI services in a self-reinforcing cycle. Similarly, Amazon applies AI to extract valuable insights from external seller data, which it leverages to its advantage. Using recommendation engines is another instance where AI is used to enhance customer lock-in across e-commerce, streaming, and app stores. Furthermore, incumbents use AI for pricing, ad targeting, and logistics, enhancing their dominance across every line of business. The strategic use of AI by dominant players in the industry reinforces their position, making it increasingly difficult for new entrants to gain a foothold and challenge established players. Consequently, the AI industry is moving toward a market structure where the majority of market share is concentrated in the hands of a few dominant players, with smaller players relegated to niche roles.

The AI industry is characterized by a concentration of market share in the hands of a few dominant players. This dominance is reinforced by the fact that upstarts lack the data advantages necessary to replicate such precise AI systems [247]. Furthermore, investing heavily in internal AI capabilities raises entry barriers and increases economies of scale for established giants, further consolidating their position in the market. The firms leading in today's AI race appear poised to further entrench their positions through skilful AI adoption across different domains. Consequently, anti-trust action may be necessary to check unfair dominance and promote competition in the market, enabling new entrants to challenge established players and fostering innovation in the industry. Technology advances and digitalization have played a significant role in explaining the divergence between productivity and typical worker compensation in recent decades. However, these gains have primarily accrued to corporate profits and share prices, with median wages lagging. Similarly, AI productivity enhancements predominantly benefit shareholders and executives rather than average workers or consumers through higher pay. Adopting automation also puts downward pressure on wages and employment levels for certain occupations, further exacerbating income inequality. Therefore, labour policies such as profit-sharing, minimum wage hikes, and stronger collective bargaining power are necessary to ensure that workers share in productivity increases enabled by AI. The distribution of the benefits of AI innovations is a critical issue that policymakers need to address if they want to create an inclusive and equitable AI landscape that benefits all stakeholders.

Adopting automation in various sectors has been observed to have a detrimental impact on certain occupations' wages and employment levels. As a result, the income gains from Artificial Intelligence (AI) tend to be concentrated among corporate owners and investors. To address this issue and ensure that the workers also benefit from the productivity increases enabled by AI, it is crucial to implement labour policies like profit sharing, minimum wage hikes, and stronger collective bargaining power. It is worth noting that the economic gains from innovative AI technologies will have varying effects on different regions. The existing technology hubs such as Silicon Valley, Seattle, Manhattan, and urban China tend to attract the majority of investment and talent, while the legacy manufacturing centres and rural areas that lack exposure to growth opportunities face concentrated displacement risks. Technology penetration also correlates with higher wage growth [247]. In light of these observations, it is imperative to ensure that the prosperity generated by AI innovations is distributed equitably rather than resisting technological change altogether. By implementing policies that support workers in adapting to these new technologies, we can ensure that everyone benefits from the positive impacts of AI. To address the divergence in regional inequality caused by the concentration of AI-related technological growth in specific regions, there is a need for place-based policies that foster tech clusters in new regions. Additionally, there is a need to adjust immigration policies to ensure that high-skilled talent is spread more widely across different regions. Promoting remote work and improving connectivity in less dense areas can also balance access to these opportunities. Although regional inequality is a complex problem, it is not inevitable and can be addressed through deliberate efforts to build inclusive tech and AI ecosystems. This requires policies that benefit the left-behind populations and ensure they can access the benefits of technological advancements. These steps make it possible to create a more equitable and inclusive society that benefits everyone.

4.1.3   Unequal Geographic Impact

The current wave of automation advances has caused a significant threat to historic manufacturing hubs, as they are at a higher risk of enduring economic instability. The implementation of industrial robotics, 3D printing, and AI-powered predictive maintenance has enabled firms to significantly reduce the need for labour [250]. Consequently, this has led to a decrease in employment opportunities as well as wages in the affected areas. Regions with above-average manufacturing concentration, such as Detroit, Cleveland, Pittsburgh, and Peoria, are already experiencing a decline in jobs and incomes, which can be attributed to the adoption of technology and globalization. The aforementioned technological advancements have played a pivotal role in reshaping the economic landscape of these areas, leading to serious implications for the local workforce and the broader economy. The continued expansion of automation technologies has created a bleak outlook for distressed regions that lack alternative industries. The once-thriving factory towns now face spiralling blight and urban decay, leading to a contraction of tax bases and public services, further accelerating the decline [251]. Unfortunately, workers displaced from automatable routine tasks also encounter difficulties finding alternative employment locally. To reverse this trend, the revitalization of these fading manufacturing zones requires vocational retraining programs, the demolition of excess capacity, the attraction of tech-oriented start-ups, and the upgrading of the urban quality of life. However, significant public-private investment and economic development expertise is imperative to achieve inclusive prosperity in these regions.

The rise of artificial intelligence (AI) in the agricultural industry has created uncertain prospects for farming communities. With the introduction of agricultural robots, predictive data analytics, and vision-enabled harvesting systems, farms can now be operated with fewer human workers [250]. The impact of automation, coupled with the existing issue of youth drain, is likely to decimate the remaining agricultural employment in rural areas, rendering livelihoods dependent on small family farms unviable, leading to the closure of storefronts and a drying up of economic activity across farm towns. Despite the fact that corporate agriculture conglomerates hold the necessary resources to invest in AI systems that benefit from efficiency gains and expanded scale [252], they employ the fewest actual farmers per acreage farmed. Thus, renewed anti-trust efforts may be warranted to prevent monopolistic outcomes. Furthermore, government programs that assist rural communities with economic diversification beyond agriculture are also essential as technology continues to transform traditional livelihoods. It is imperative to implement anti-trust policies and government programs that promote economic diversification to mitigate the negative impact of automation on traditional livelihoods. As a result of recent advancements in AI technology, the need for physical proximity and local service knowledge has been significantly reduced. With the introduction of autonomous delivery drones and remote expert advice, even small-town businesses such as grocers, bankers, insurers, lawyers, and accountants are at risk of being displaced by online AI-powered platforms [250]. In the past, big box retail stores have drawn business away from small, local shops. Now, the removal of localization advantages by AI-powered platforms may result in the closure of remaining mom-and-pop businesses.

In order to revive rural communities and small towns that have been hollowed out, there is a need to reinvent their economic purpose achieved by transitioning to industries such as artisanal manufacturing, agri-tourism, clean energy, and retirement community services, which have the potential to provide new directions for these communities. Additionally, advanced connectivity infrastructure can unlock remote work opportunities, thereby expanding the scope of employment. However, it is important to acknowledge the realistic limitations of locales restricted by permanent population loss or geographical isolation. It may be necessary to consider the consolidation of struggling settlements in order to prevent blight in such cases. The consequences of technology and globalization on certain regions can be severe, resulting in a combination of job losses and small business closures that impose immense economic strains on the affected areas. As a result, household incomes often decline precipitously, and the limited reemployment options available can lead to mortgage defaults and foreclosures, leading to disinvestment and rapid deterioration of neighbourhoods [250]. Despite the declining property values and rents, maintenance costs for properties often remain high, leading to abandonment. Abandoned properties are often a magnet for crime and create potential safety hazards. This downward spiral is often exacerbated by the inability of cash-strapped city governments to provide adequate services or raze derelict properties, resulting in a loss of tax revenues, leading to bankruptcy, further depressing the prospects of the region. As once-vibrant areas empty, many remaining residents fall into poverty without the infrastructure necessary for upward mobility. To avoid such destitution, planning for regional transitions in advance rather than reacting only after vicious cycles have taken hold is essential. Rural communities are often disadvantaged in an AI economy due to their geographic remoteness from urban economic centres. Compared to cities and suburbs, sparsely populated areas have fewer professional and technical job opportunities, making it difficult for rural residents to find employment in their fields. Additionally, the adoption of technology tends to automate traditional rural occupations in agriculture, mining, forestry, and utilities [250]. Furthermore, minimal internet connectivity obstructs remote work and educational opportunities, which can be limiting for those who prefer to live in rural areas. Start-up formation in rural areas also tends to lag behind that of urban centres. Younger populations often migrate to metropolitan areas, leaving behind an ageing population with higher dependency costs. The declining economic prospects of rural areas can discourage immigration, making it difficult to revitalize small towns. Delivery of government services becomes more expensive per capita with diffuse populations. Lagging infrastructure like roads, broadband, and housing also hampers attracting new industries. The turnaround for rural communities depends on upgrades to connectivity and infrastructure, vocational training, and cultural shifts that value rural lifestyles.

When it comes to the economic impact of AI, struggling traditional heartland regions seem to be at a disadvantage as compared to major coastal cities and tech hubs. Cities such as Silicon Valley, Seattle, New York, and Boston already account for a disproportionately high share of high-tech talent, venture capital, and innovation, as noted by Moretti [253]. The correlation between tech permeation and wage growth, home values, and other prosperity indicators is also quite strong. Urban density creates positive feedback loops that attract top talent and capital, further enhancing the economic prospects of these cities. Some experts predict that a barbell economy could emerge, with most jobs and wealth being polarized between sophisticated megacities and the rest of the country [254]. Historically, new technologies have eventually spread economic benefits more widely through productivity gains. However, this process could take generations without proactive policies accelerating tech diffusion. Governments need to play an important role in providing transitional assistance and opportunities to communities that may be negatively affected in the short term. Such policies can facilitate a more equitable distribution of economic benefits and promote a more inclusive society. The growing divergence in economic fortunes between metropolitan areas and heartland regions poses a significant risk to the economy and threatens to stoke political and social divides. Rural voters already perceive growing stratification between globalized urban elites and the traditional working class, highlighting the need for policies that promote more equitable distribution of economic benefits [255]. As automation continues to advance, it threatens to hollow out remaining middle-class manufacturing and white-collar jobs in regional mid-tier cities. Furthermore, the concentration of technology wealth in big cities like New York and other tech hubs worldwide exacerbates inequality and geographic sorting by class. Populist resentment stems partly from stagnant interior regions being left behind by tech booms. To curb such backlash, providing transition assistance for displaced workers and distressed areas is essential. However, efforts to reduce megacity housing costs, connect remote work, and foster cultural integration can also help diversify access to high-tech gains. A more inclusive society recognises and addresses the growing economic disparities between different regions and takes proactive steps to narrow the gap.

The uneven geographic impacts of AI and automation make it clear that locally targeted policy interventions, not just broad national measures, will be essential in addressing the challenges faced by different regions. Regional economic development strategies that are tailored to local strengths and assets provide useful templates for addressing these challenges [256]. Sector-specific worker retraining programs are another effective measure that can equip affected populations with the skills necessary to ensure occupational mobility. Tax incentives can also encourage new industries to locate in hollowed-out zones, promoting economic development in underdeveloped regions. Preparing the next generation with relevant skills and connections will also prevent brain drain outflows, ensuring that local regions have a steady supply of skilled workers. Fibre broadband buildouts can open up new digital opportunities and facilitate economic growth in these regions. Planning and investing continuously in local transitions can prevent severe shocks and provide displaced workers with the necessary support. While national policies can set the direction, empowering communities to adapt AI technologies for inclusive growth remains vital. By doing so, local regions can ensure that they are well-positioned to benefit from AI's opportunities while mitigating its negative impacts.

4.1.4 *Marginalization of Disadvantaged Groups*

Algorithmic systems, which are developed based on historical data, may perpetuate and amplify existing gender and racial biases. This issue arises due to the fact that word embedding models inherit human associations that are often messy and may encode problematic stereotypes [257]. Furthermore, image recognition datasets tend to under-represent women and minorities, leading to higher error rates for these groups. In addition, predictive policing models tend to over-flag low-income minority areas based on past over-policing practices [258]. These illustrate the extent to which machine learning models can absorb social prejudices and inequitable power dynamics from the datasets on which they are trained. Unless monitored and corrected appropriately, AI has the potential to automate and exacerbate discrimination on a large scale that goes beyond human capacities. Therefore, mitigating algorithmic bias requires diversifying data collection, proactively engineering fairness criteria, and centring impacted voices throughout the machine learning pipeline. These measures are essential to ensure that AI systems are developed and utilized in a fair and just manner. The biases against disadvantaged groups in AI systems can be attributed to several factors, including data limitations and narrowly optimized evaluation metrics. It is important to note that machine learning models are only as good as the data they are fed. For instance, image and speech recognition systems trained on demographically homogeneous data will likely suffer from performance gaps across populations. Additionally, crime prediction models tend to overfit correlations from historically skewed police datasets, leading to biased outcomes [258]. The use of static and limited test sets to benchmark progress also incentivises algorithms to excel on mainstream data at the expense of marginalized groups. Furthermore, social media platforms prioritising engagement have learned to maximize outrage and polarization, further exacerbating existing social divides [259]. To prevent such exclusions encoded performatively into the models, adopting a rights-based AI design that prioritizes representative data collection and inclusive testing is crucial. By doing so, the resulting AI systems can be designed to be more equitable and just, promoting social justice and fairness for all.

The lack of transparency in many commercial AI systems poses a major challenge when it comes to identifying and addressing issues of discrimination. The proprietary black box nature of these systems makes it difficult for independent auditors to assess the presence of discriminatory practices. In many cases, developers themselves may not fully understand the modelled connections inside multilayer neural networks, making it even more challenging to identify problematic biases [181]. The opaque nature of these systems can enable questionable or even illegal practices to go unchecked in various applications, such as hiring, lending, facial analysis, and predictive policing. To address these issues, AI researchers advocate for algorithms to be inspectable, interpretable, and accountable rather than being treated as inscrutable black boxes. Models that impact public welfare must be available for bias testing by external auditors. Governments also have a role to play in mandating algorithmic transparency, similar to financial disclosures. Without visibility into the inner workings of AI systems, ensuring fair treatment of vulnerable populations remains impossible. Therefore, it is essential to promote transparency and accountability in the development and deployment of AI systems, particularly when addressing discrimination issues and ensuring equitable outcomes for all. The deployment of unconstrained AI on centralized platforms presents unique challenges in reproducing historical discrimination patterns at an unprecedented scale and speed. Machine learning models are designed to mechanically extrapolate correlations and optimize goals without inherent justice or social equity considerations. This lack of thoughtfulness in design can result in AI systems inheriting humanity's past prejudices and amplifying the existing ingrained social hierarchies [260]. For instance, résumé screening algorithms that are trained on companies' past applicant records can embed legacies of bias in hiring and promotion. Similarly, loan risk scoring models based on historical datasets tend to bake past denials against minorities into automated decisions.

To neutralize unfair biases, it is crucial to firewall off tainted training data and emphasize disparate impact assessments over pure predictive accuracy. Furthermore, the scale of AI deployment necessitates thoughtful governance to ensure that AI systems are designed and deployed in a fair and just manner. It is important to note that AI systems are not neutral and can perpetuate harmful biases if not designed thoughtfully. Therefore, it is essential to promote diversity and inclusivity in data collection and prioritize fairness and justice in algorithmic decision-making to ensure that AI systems are designed and utilized in a fair and just manner. Although AI can exceed human intelligence on narrow cognitive tasks, it is important to acknowledge that algorithms lack the capacities for emotional nuance, ethics, and social contextualization. As a result, final high-stakes decisions on welfare, justice, and personal treatment should not be fully ceded to statistical models. Human oversight is needed to balance the inherent limitations of AI and to ensure that complex decisions are made in a fair, just, and ethical manner. However, it is important to note that disadvantaged groups often lack agency in contesting dubious automated judgments, further highlighting the need for ethical considerations and human oversight in developing and deploying AI systems. Structuring hybrid systems that leverage human and machine intelligence can be an effective strategy for counteracting biases in AI. Judges, loan officers, and healthcare workers should treat computational outputs as advisories to enhance, rather than replace, specialized expertise and situational judgment [246]. By centring on human dignity and individual context, it is possible to guard against marginalized groups being dehumanized by insensitive automation. In this way, hybrid systems can provide the benefits of both human and machine intelligence while also minimizing the risks of biased decision-making. Moreover, by incorporating diverse perspectives and situational context into the decision-making process, it is possible to ensure that AI systems are designed and utilized in a fair and just manner. Ultimately, such an approach can help to promote social equity and justice while also realizing the potential benefits of AI for society as a whole.

As AI systems become increasingly prevalent in the workforce, preparing workers for complementary collaboration with these systems, rather than displacement by them, is essential. This requires broadening access to digital skills training and STEM education, which can be a challenge for disadvantaged students from underfunded schools which often lack the foundational skills needed to pursue tech careers. Furthermore, retraining programs tend to favour highly educated incumbents over those needing upgrades. Initiatives such as LabourTech partnerships, apprenticeships, and vocational IT tracks can help to close readiness divides and ensure that workers are equipped with the skills needed to thrive amidst automation transformations [261]. Proactive policies that broaden science, technology, and lifelong learning opportunities can empower citizens to succeed in the digital age. Governments can play a key role in incentivizing educational institutions to prioritize inclusion and flexibility while promoting diversity in tech careers and usage as a product feature rather than an afterthought. It is important to note that the failure to democratize access to digital skills training and STEM education can cede the economic gains of tech to existing elites, further exacerbating existing social inequalities. Therefore, it is crucial to prioritize diversity and inclusion in developing and deploying AI systems while promoting equal access to digital skills training and STEM education to ensure that all workers can benefit from AI's opportunities. The data-driven targeting enabled by AI has the potential to amplify existing barriers to economic mobility for historically marginalized groups. When layered on top of unequal education and housing systems, supposedly neutral algorithmic decision-making can calcify old divisions, further entrenching occupational segregation and reflecting past discrimination in financial services [200]. Despite noble intentions, AI applications built hastily atop flawed historical datasets inevitably amplify social stratification. To counter these patterns, assessing AI systems' disparate impacts on underprivileged populations, rather than just focusing on aggregate efficiency, is crucial. Technological advances must be met with economic policy and corporate practices prioritising lifting up those previously left behind. This requires a concerted effort to promote diversity and inclusivity in the development and deployment of AI systems while ensuring that they are designed to promote economic mobility and social equity for all. Addressing these challenges requires a multifaceted approach that includes technological and policy solutions. By centering the needs and experiences of historically marginalized groups in the development and deployment of AI systems, it is possible to ensure that these technologies are utilized in a fair and just manner, promoting social equity and economic mobility for all. The economic divide between prosperous regions and struggling locales is at risk of widening in an AI economy without conscious policy. This is because automation advantages tend to cluster around existing tech hubs, while rural areas and former manufacturing centers face high displacement risks from technology adoption [262]. As a result, new opportunities tend to concentrate around cities that are attracting AI specialization. To prevent further exclusion of already marginalized populations, governments need to target transition programs, infrastructure investment, and economic development incentives towards vulnerable communities that are being disrupted by technological change. Firms that are leading advances, such as self-driving vehicles, also have corporate social responsibility duties to their historical manufacturing workforces. Concerted efforts at inclusion are necessary to temper the risks of AI exacerbating regional inequality. In this way, it is possible to ensure that AI is developed and utilized to promote social equity and economic mobility for all while minimizing the risks of exclusion and marginalization of vulnerable communities. By targeting policy interventions towards vulnerable communities, promoting corporate social responsibility, and prioritizing inclusive economic development, AI's potential benefits can be harnessed while minimizing its risks.

4.1.5 Policy Approaches to Ensure Inclusive Prosperity

In today's world, technological advancements have led to the development of artificial intelligence (AI) and automation, which have the potential to enhance productivity and economic growth significantly. However, these technologies also threaten to exacerbate inequality, unemployment, and disruption for many workers [263]. In order to promote inclusive prosperity, policymakers need to adopt intelligent regulations, incentives, and investments. As suggested by leading experts in the field, effective policy measures include allocating resources to retrain the workforce, revising educational curriculums, providing tax incentives to encourage responsible technology adoption, mandating algorithmic transparency, promoting diversity in AI, establishing public-private partnerships, rethinking social safety nets, considering universal basic income, updating anti-trust laws, and coordinating global AI governance, as outlined by [264]. The advent of artificial intelligence (AI) and automation is set to revolutionize job markets, leading to the contraction or elimination of some human roles and tasks. While these developments are expected to create new and better job opportunities, many incumbent workers may face the risk of skills obsolescence and unemployment [242]. In order to ensure that displaced workers can transition equitably, governments must allocate resources towards funding robust retraining and "upskilling" programs. According to the Organization for Economic Cooperation and Development (OECD) [265], over 140 million global workers must be reskilled by 2030, necessitating substantial public and private investment. Specific policy recommendations include providing tax credits or grants to employers for approved retraining programs, fully funding community and technical colleges to offer relevant certification programs, and reforming unemployment insurance to allow enrollment in transition training during periods of joblessness [266]. Retraining programs must target vulnerable groups, such as less-educated, low-income, and minority workers, who are disproportionately likely to be impacted by automation, as emphasized by Brussevich et al. [267]. Governments can expand Trade Adjustment Assistance to support workers displaced by technological advancements, not just globalization [268]. Large-scale training programs that partner with employers can efficiently reskill at-risk workers with proper funding and support from the government, ensuring that the transitions driven by AI are smooth and inclusive. In addition, policymakers must prepare students and future workforces with a balance of adaptable human skills and job-ready technical skills essential for the AI economy, as highlighted by the World Economic Forum [269]. Education policies should strike a balance between immersive technology exposure and "human-centric" capacities, such as creativity, communication, problem-solving, and socio-emotional intelligence, which are highly valued in the AI age [270].

To foster the development of adaptable human skills, primary and secondary curriculums need to be updated; core courses should emphasize computational thinking, collaborative projects, and real-world problem-solving to equip students with the necessary skills for the AI economy. Governments can play a role in this by allocating resources towards teacher training and developing openly licensed digital course content to facilitate rapid integration, as suggested by UN-

ESCO (2021). Making computer science and data literacy universal graduation requirements can also help boost readiness. In addition, higher education policies play a complementary role. Governments should incentivize colleges and technical schools to expand AI and technology-focused programs through funding and favourable student loan policies which are of high significance [271]. Tax incentives and public-private partnerships can also provide students with early exposure to tech internships and careers. With students and future workers adequately prepared, countries can more equitably distribute the economic gains of AI. Although rapid AI and automation adoption can enhance productivity, it poses a risk of economic exclusion and turbulence if human workers are displaced without sufficient support. Tax incentives are a useful policy tool to promote responsible tech uptake [262]. For instance, as recommended, tax credits could be offered to employers who provide displaced workers with severance, transition benefits, and funded retraining opportunities. Another approach could be to impose tax penalties on rapid automation without commensurate workforce investments. Different payroll tax rates could also incentivize employers to augment roles with technology while ensuring that workers' earnings remain unaffected [272]. For example, lower tax rates would apply to workers provided with productivity-enhancing tools compared to workers displaced outright. Tax incentives can also be used to reward businesses that implement algorithmic transparency, diversity hiring initiatives, and other responsible AI practices, as highlighted by Floridi et al. [273]. Well-designed tax policies can encourage equitable tech adoption and promote an inclusive prosperity dividend.

Governments must mandate transparency standards to ensure that AI remains aligned with societal values. Specifically, providers of algorithmic tools in areas such as hiring, lending, policing, and beyond should be legally required to audit systems for issues such as bias and submit explainability reports [274]. The system documentation should provide details about the data used to train algorithms, the benchmarks and tests performed, the steps taken to address unfair outputs, and the procedures to follow in case of disputes over algorithm functionality or determinations and externally conducted audits could also provide independent assurance. The EU's proposed Artificial Intelligence Act [275] is a leading model for such oversight, demanding that algorithm providers assess risks and provide transparency per proportionality principles, with violations incurring stiff financial penalties. Experts concur that similar audit and explainability laws are needed in other regions to monitor AI systems, as recommended [276], enhancing public trust and accountability around automation. With transparency, citizens can better ensure that technologies align with inclusive values while promoting shared prosperity. Utilizing homogenous teams and unrepresentative data for developing algorithms poses a significant risk of replicating and amplifying unfair societal biases, as noted by West et al. (2019). Therefore, governments must implement policies and programs that promote diversity and inclusion in institutions involved in researching, designing, and deploying AI systems. Such measures may include funding opportunities, procurement standards, hiring mandates, and anti-discrimination laws that can help ensure AI systems reflect diverse perspectives and values [274]. For instance, governments may favour collaborative teams that span gender, racial, and socioeconomic diversity by allocating research grants preferentially to them. Similarly, policymakers can make diversity representation a condition of government procurement contracts for commercial AI system providers or require corporate board diversity for favoured tax treatment [273]. In addition, strict anti-discrimination laws need to be enforced to ensure that individuals who have been historically underrepresented in tech, such as women, minorities, disabled individuals and non-conforming genders, enjoy equal opportunities in AI roles. With inclusive participation, the economic benefits of AI can be shared more broadly across all groups, thereby promoting a fair and equitable society. The achievement of inclusive prosperity through the use of AI and automation necessitates active collaboration among industry practitioners, government policymakers, and academic researchers from diverse fields such as computer science, law, ethics, and economics [264]. Each stakeholder brings unique capabilities towards implementing emerging technologies, anticipating potential risks, and enacting evidence-based policy responses. Therefore, multi-stakeholder partnerships should be formed to address challenges such as job market turbulence and algorithmic bias and to ensure accessible digital infrastructure to enable equitable distribution of benefits from innovations. Specifically, such partnerships should focus on developing effective solutions to these challenges, promoting innovation and technological advancements that are inclusive, socially responsible, and sustainable. This requires formulating policies that promote equal opportunities and equitable access to resources for all communities, especially those that are marginalized or underrepresented. By working together, stakeholders can leverage their respective strengths to create a shared vision of an inclusive and prosperous future enabled by AI and automation.

There are promising examples of cross-sector partnerships that have been established to address the challenges posed by AI integration in the economy and society, such as Germany's Platform Lernende Systeme, Japan's Artificial Intelligence Technology Strategy Council, and Canada's Pan-Canadian AI Strategy [274]. These partnerships aim to coordinate national plans focused on fostering cutting-edge research and development, developing commercialization pathways for innovations, establishing standards for responsible implementation, and building skills to enable workforces to transition fluidly amid technological shifts[277]. By collaboratively planning for an AI-integrated economy and society, stakeholders can help steer automation's productivity towards a more broadly shared prosperity. Furthermore, these partnerships can facilitate the development and implementation of responsible AI technologies that promote equitable access to opportunities and resources for all members of society, thereby enabling the realization of a sustainable and inclusive future. The utilization of AI is known to increase productivity; however, there is a risk of augmenting inequality if the existing systems fail to support those who are economically marginalized. As per policy experts, the disruption in the labour market caused by AI and automation will necessitate the reconstruction of income supports and social safety nets to ensure that households have sufficient livable incomes [242]. Suggestions range from making amendments to the current systems, such as disability payments and unemployment insurance, to more extensive measures, such as introducing an unconditional, universal basic income (UBI) or guaranteed job programs [278]. Incremental reforms could also involve increasing benefit amounts, expanding unemployment eligibility periods, consolidating various programs for simplified access, and creating funding pools to support those who have been displaced by technology specifically [279]. UBI pilots and proposals, which provide regular income to all citizens or households irrespective of their working status, have been suggested as a more accessible and less stigmatizing alternative to complicated, conditional welfare systems for the AI era. However, the affordability of UBI and its impact on workforce participation remain topics of debate. Hybrid proposals such as universal basic services funding essential needs such as healthcare, transportation, housing, and food also hold the potential for updating social safety nets for an AI-driven world [280].

However, some critics challenge the feasibility of UBI's funding and its effectiveness in reducing poverty compared to the existing targeted programs [281]. Additionally, some argue that basic income may diminish the motivation to work or pursue education and skills training, thereby worsening inequality in the context of AI automation instead of mitigating its impact [282]. While the evidence remains limited and debated, experts concur that pilot experiments can provide valuable insights into the optimal policy design for UBI, its impact on labour markets, and distributional effects on lower-income populations [283]. Carefully planned pilots will assist policymakers in making informed decisions regarding the feasibility of UBI in conjunction with other proposals for updating social safety nets in a changing economy. The use of AI and platform economics has the potential to stimulate innovation and productivity. However, it also poses a risk of excessive corporate power and data concentration, enabling firms to exploit their dominant market positions or access user data . In order to maintain market dynamism and protect consumer interests, governments must update their anti-trust policies and enforcement. Such updates may include lowering evidentiary thresholds to demonstrate anti-competitive harms and acknowledging that network effects, accumulated data, and algorithms provide tech platforms with significant leverage to establish dominance and limit consumer choice [284]. Strengthening merger oversight aims to safeguard future competition even in cases where immediate price gouging is not evident [285]. Strict data interoperability and portability rules can also reduce user lock-in effects that favour incumbents. Overall, the modernization of century-old anti-trust policies acknowledges that the use of

AI-enabled economies necessitates greater vigilance against excessive corporate power to ensure the equitable distribution of prosperity.

As countries continue to implement national policies on funding, education, incentives, transparency, and competition around AI, experts have emphasized the need for international coordination due to the borderless nature of emerging technologies [286]. Collaboration is needed to set risk-based rules on research areas such as lethal autonomous weapons. Transnational efforts would also enable the sharing best practices on topics such as data governance frameworks that promote innovation while respecting user rights and testing social welfare policies like basic income schemes amidst rising automation. Further, calls exist for the establishment of a Global AI Coordination Council, similar to the World Trade Organization overseeing global trade, to establish shared principles for AI usage that advance human rights and well-being worldwide [287]. With aligned international AI governance, countries can implement emerging technologies with sensitivity to local contexts while working together towards the highest shared values and prosperity. The increasing capabilities of AI and automation hold immense potential to boost global prosperity. However, it also poses risks such as worsening inequality, job displacement, and concentration of power during the transition. Pre-emptive policy interventions across funding, incentives, education, transparency, competition, and domestic and global coordination provide tools to steer technological change towards broadly shared prosperity. Leaders must enact inclusive policies that balance accelerating progress with easing transitional downsides so that the immense possibilities of AI serve all people.

4.2   Redefining Success - Shifting Cultural Norms and Values with Changing Work Dynamics

4.2.1   Rethinking Productivity and Performance

The impact of artificial intelligence (AI) and automation technologies on work across various industries is profound and transformative. With the takeover of routine and repetitive tasks by AI systems, human strengths in areas such as creativity, empathy, and collaboration are becoming increasingly crucial in evolving work contexts [288]. Nevertheless, the prevalent constructs around productivity and performance, which are rooted in quantifying time spent on tasks, often fail to appreciate the multifaceted human capabilities. Thus, rethinking outdated paradigms is essential to motivate human-AI collaboration that targets quality, innovation, and societal impacts rather than just efficiency. In many workplace cultures, assessments of productivity and value are rigidly based on time, such as hours worked, shifts logged, and Facetime demonstrating dedication. However, these temporal metrics become less relevant as AI automation takes over routine tasks and alters job designs. When predictive algorithms and exoskeletons assist human judgment in executing intricate work, overall contributions better reflect hybrid potential than time on task. Experts suggest adopting metrics that evaluate workers more on the end value produced through both automated and augmented processes rather than input-focused efficiency. Instead of counting hours logged, the evaluation would shift to the quality, innovation, and stakeholder impact of overall contributions [289]. For example, as AI chatbots handle customer inquiries, human capacities building trust through nuanced interactions become a priority metric. It is crucial to deliberate whether existing metrics serve goals amid changing workplace contexts before redesigning for the future [290]. The transition to a workplace where AI and humans collaborate poses challenges to ingrained constructs. Contribution-centered assessments better capture both human strengths and AI's scale in solving problems. With adjusted norms, workers can focus their energies on the most value-adding challenges without redundancy fears. Being open to reframing metrics around hybrid human-AI outputs promises both productivity and worker satisfaction gains. Unlike routine tasks that are readily automated, unique human strengths such as creativity, complex communication, and contextual reasoning remain unmatched by machines [291]. It is argued that norms and assessments should especially value these irreducible contributions that AI cannot replicate, thus motivating human energies towards augmented challenges that best leverage multidimensional capacities.

As algorithms handle routine work, metrics should increasingly reward innovation, experimentation, abstract strategy, and imaginative pursuits that reveal human potential. Likewise, collaborative, trust-centered relationships and emotional intelligence showcase humans' contextual mastery [289]. With updated metrics that judge these irreplaceable contributions, workers can complement AI efficiencies with transcendent vision and meaning-making for superior results. Education models promoting design thinking and humanities training can help develop these multifaceted areas [292]. Rotational programs can build empathy across business and cultural roles. Forgiving environments that celebrate managed failures can accelerate learning. Adjusting norms and assessments to motivate creative leveraging of AI strengths with expansive human collaboration and exploration promises both productivity and activation of deeper human talents. Human skills in relationship-building, cultural understanding, and empathy remain distinct from artificial intelligence technologies' prowess in technical optimization and computation [293]. However, workplace cultures that are narrowly focused on rote efficiency risk underappreciating these invaluable emotional, social, and cultural intelligences that are nurtured through lived experience. Updating paradigms should clearly recognize that AI complements but cannot substitute for humans' capacities in judging subtle social cues, respectfully navigating diversity, relating ethically, and the emotional resonance that drives trust. Emphasizing humanities training and rotating staff across business and cultural roles can strengthen appreciation for these entire intelligences. Normalizing empathy, compassion, diversity, and sensitivity as fundamental workplace skills, not "soft skills" ancillary to technical capabilities, appropriately rewards care, ethics, and moral imagination that AI systems lack [289]. Allowing sabbaticals for emotional growth signals valuing multifaceted social richness. Balanced human+machine collaboration hinges on assessments that nurture uniquely human strengths that machines lack. Prevalent workplace productivity constructs centred around rote efficiency, such as task volumes, turnaround times, and output scales, risk growing outdated in an AI-powered context. Automation now rapidly handles high-volume repetitive tasks while augmenting technologies like predictive analytics and exoskeletons expand human capabilities in executing intricate work. These shifts warrant rebalancing norms that judge overall work quality, innovation, and stakeholder impact as priorities rather than just quantified speed or scale, which algorithms excel at [294]. Updated paradigms should highlight irreplaceable human contributions - whether intricate strategies or ethical relationships - that AI cannot replicate. Productivity Constructs should especially reward manifold thinking, cross-cultural fluency, imaginative problem-solving, and other multidimensional impacts that are resistant to data quantification—tracking quantitative efficiency alongside judging hybrid potential. With norms evolving to value complementary abilities, integrated human-machine teams can enhance productivity and meaning amid automation. Solely efficiency-focused productivity paradigms inadequately capture multifaceted human potential and integrated human-AI capabilities in augmented contemporary workplaces. Mature leaders should rethink outdated cultural norms and assessments to especially judge irreplaceable human strengths around creativity, ethics, emotional intelligence, design, and strategy - the talents no algorithm can replicate. Updated constructs valuing both systemized and social outputs empower humans to leverage automation for productivity and purpose. Rethinking norms to unlock hybrid intelligence promises superior prosperity.

4.2.2 Changing Workplace Culture and Dynamics

The amalgamation of artificial intelligence (AI) and versatile automation technologies significantly transforms the correlation between humans and work. With algorithms taking over repetitive tasks, the importance of distinct human abilities such as strategy, creativity, empathy, and collaboration is amplified in augmented roles. Consequently, it is crucial to update workplace cultures to empower human talents [291]. Leading research suggests that organizations should adopt flatter, decentralized structures, offer more flexible schedules, renew emphasis on purpose, and provide continuous support for employee growth in response to the AI-driven competitive landscape [295]. Failure to revise entrenched mindsets may result in disengaged workforces, innovation setbacks, and loss of top talent to forward-thinking competitors. The conventional workplace frameworks, characterized by multi-layered hierarchical organizations and centralized authoritarian decision-making processes, were established to coordinate large workforces performing manual and repetitive tasks at a massive industrial scale. However, the integration of AI, which is capable of handling rote work and analysis, necessitates a flatter, decentralized coordination mechanism that empowers employees' irreplaceable judgment. By eliminating excessive bureaucracy and devolving autonomy around priorities, processes and resource utilization, organizations can better leverage human strengths such as creativity, relationships, and market insights that rigid control structures often constrain. In practical terms, the process of delayering middle management helps flatten information flows and enhances access for faster adaptation [296]. The formation of cross-functional teams equipped with decentralized budgets and decision-making rights allows the piloting of experiments, knowledge sharing, and trust-building through lateral coordination. Furthermore, regular peer feedback and participative goal-setting complement top-down performance reviews. Updated rhythms, such as all-hands meetings, hackathons, and open critiques, expand both transparency and idea flows. With the automation of mundane tasks, decentralized and networked team architectures can help focus uniquely human energies towards higher judgment, innovation, and meaningful work. Location-bound and time-regimented work paradigms appear to be outdated in the face of growing workforce mobility and ubiquitous connectivity. Employees increasingly seek and maintain high productivity levels through asynchronous virtual work facilitated by collaborative technologies [297]. Younger talent, in particular, expects flexibility that accommodates individual circumstances, health, and family needs, which are impossible to meet within rigid office presentism constraints. Embracing updated norms that support remote work can attract and empower contemporary talent. Options that accommodate flexible hours, condensed weeks, seasonal rhythms, or periodic sabbaticals also demonstrate a commitment to employee well-being and trust. Research confirms that workers granted flexibility balance productivity with their circumstances without abusing their freedoms [298]. Managers should focus more on the overall outcomes achieved than physical oversight. Securing buy-in around shared purpose and providing collaborative tools suites can smooth adaptation pains as organizational cultures transition from standardized schedules towards supported customization that meets both human needs and business deliverables. In environments that are optimized for rote efficiency, work can often feel dehumanized and reduced to soulless drudgery. However, automation has freed employees from these mundane tasks, and they now crave purpose and passion in their augmented, creative roles. In this context, updated management norms should strike a balance between fiduciary responsibilities and energizing workers around meaningful challenges that leverage their uniquely human strengths. Rewarding social impacts, creative risk-taking, and community building can foster cultures where technical optimization supports, rather than supplants, reflective purpose. Conducting regular meaning check-ins, defining individual purpose, and mapping job alignment can help prevent disengagement more effectively than strict monitoring alone. Additionally, initiatives such as customer advisory panels and non-profit partnerships can strengthen organizational commitment to societal values. Accommodating voluntary service programs can demonstrate institutional values beyond financials. With automation increasingly handling productivity, humans who master strategy and relationships should find motivation in problems that require not just skills but wisdom. Updated management norms that balance economic and social value can sustain fulfilment amid workflows that are increasingly shaped by artificial intelligence capabilities.

Given the accelerating technological turnover of skills, continuous learning that is in step with market evolution is crucial. Employers should robustly support regular staff education that accommodates both career development needs and the adaptation to changing workplace contexts [299]. Tuition assistance, stipends, and time allowances can enable enrollment in both degree and microcredential programs, maintaining competitiveness. Rotational cross-training can prevent siloes as teams integrate new solutions. Expert-led seminars, virtual reality simulations, and remote access to international colleagues can accelerate perspective shifts. Curating an always-learning organizational culture begins with messaging continuous development as a shared imperative from executives to the frontline staff. Managers should guide their staff in aligning skills development to emerging roles, not just task needs. With lifelong learning becoming the new norm, supporting self-driven employee education both economically and culturally conveys an enduring commitment to human capital development amid workplace transformations. Forward-looking leaders should balance efficient technical infrastructure with updated environments that are attuned to human needs around autonomy, creativity, and purpose. The hyperfocus on output metrics risks burnout and disengagement without parallel input that values staff satisfaction, subjective well-being, and holistic development journeys beyond skills training. In complex operating environments, which are increasingly influenced by artificial intelligence capabilities, organizational prosperity is differentiated by unique human strengths such as judgment, relationships, empathy, and wisdom. However, prevailing workplace constructs often constrain these talents with bureaucracy, control, and a singular efficiency focus. As AI increasingly handles the scale execution of routine tasks, leading firms should rethink traditional paradigms by embracing flatter teams, remote flexibility, and ongoing learning. Updating cultures that focus on decentralizing authority, refreshing purpose, and supporting multifaceted employee growth promises productivity and lasting values amid automation.

4.2.3   AI as an Opportunity to Improve Society

In addition to efficiency gains, artificial intelligence (AI) technologies offer profound opportunities to elevate human welfare, provided that governance keeps pace with technological change. The automation of mundane work liberates humans to pursue more meaningful endeavours. Channelling algorithms towards societal priorities, such as sustainability, promises new solutions to intractable challenges, as Manyika et al. [300] noted. Ensuring that AI access, oversight, and priorities serve the public good rather than solely economic interests is essential to improve livelihoods. However, thoughtfully enacted policies are crucial to promoting safety, accountability, and equitable impacts amid rapid development [301]. Today's leaders have a responsibility to collaborate to steer the responsible rise of AI, not as an inevitable tide but as a choice to uplift shared dignity.

Artificial intelligence (AI) promises immense productivity gains by automating routine physical and cognitive work. However, beyond efficiency, machines handling drudgery provide an opportunity for societal cost savings by freeing human talent to tackle society's grand challenges and higher-order passions. Teachers can gain more bandwidth for cultivating critical thinking unfettered by standardized test preparation. Doctors can focus on diagnostic intuition with assisted data review. Clerks can assist more callers with empathetic advice as bots field routine inquiries. Across contexts, AI can lighten the burden of labour from necessities towards nobler callings. Governance plays a crucial role in directing automatons to tackle dreary tasks, while updated policies support individuals' purpose-aligned transitions. Educational curriculums are increasingly emphasizing digital literacy and creative arts over rote formulas. Labour regulations, such as secondary incomes, portability schemes, and collective representations, can help smooth workforce modernization. With thoughtful coordination, AI promises to deliver both productivity and meaning to the workforce rather than precarity. By prioritizing the development of human skills and aligning AI with societal values and aspirations, technology can be harnessed to empower individuals and organizations alike, enabling them to thrive in a rapidly changing world. In addition to business efficiencies, channelling algorithmic insights to societal priorities offers immense potential for improving lives at scale. Climate researchers are deploying neural networks to predict extreme weather and catastrophic impacts, which were previously impossible to synthesize manually. Medical labs are utilizing computational diagnostics to aid in the discovery of tailored therapies. Smart urban architectures are reducing emissions by optimizing systems-level flows. By channelling AI through cross-sector open collaborations, problem-solving capacity on challenges like sustainability, which previously seemed intractable, can be multiplied [302]. This approach has the potential to transform entire industries, enabling them to operate more sustainably and efficiently while also advancing the common good.

Governing AI for social good requires updating policies, research funding, and public-private partnerships that prioritize human welfare over sole efficiency. The grant criteria should assess the social impacts of scientific contributions to guide their discovery according to pressing needs. International accords should align innovation ecosystems for the equitable sharing of breakthroughs. The flexible IP regimes should ensure access and support humanitarian applications. With technological transformation being inevitable, collective steering towards justice and empowerment promises AI raising global wisdom over simply accumulating power or wealth alone at the expense of shared interests. By working collaboratively across sectors and borders, we can harness the potential of AI to drive positive social change and empower people around the world. For AI systems that impact lives, representative data and participation behind development are crucial both ethically and functionally. Homogenous teams and biased data amplify unfair impacts across racial, gender, and socioeconomic divides, as noted by West et al. Policies promoting diversity and inclusion in research contexts enhance system security, consumer trust, and innovation by tapping into the lived experiences of presently marginalized groups. Measures such as dedicated funding, updated hiring practices, and accountability structures should ensure that AI reflects varied user values, avoiding exclusion despite the best technical intentions. By embracing diversity and promoting inclusion, AI can be developed and deployed in ways that are more equitable and responsive to the needs of all users, leading to more positive social outcomes. The implementation of inclusive policies in practice involves a range of measures such as targeted R&D grants, non-profit incubators that accelerate access, procurement contracts that require ethical accountability structures, and regulations that mandate rights protections by design, thereby preventing impersonal systems from exploiting vulnerable communities. In order to expand participation, it is necessary to update STEM pipelines and career paths, which can improve representation and cultural competence firm-wide. The promise of equitable prosperity lies in fostering a spectrum of voices that shape AI directions. The ultimate goal of AI productivity should be to enhance human development and ecological sustainability, which are the underlying prerequisites to shared prosperity [303]. However, corporate and public sector decision-making often ignore full-cost externalities and non-price social impacts. Therefore, AI governance should mandate the deliberate assessment and addressing of societal tradeoffs instead of deferring reflexively to market mechanisms alone. Forward-looking leaders can steward the updating of planning processes, balancing efficiency with ethics [304]. Several oversight mechanisms have shown promise in internalizing social welfare, ranging from algorithmic impact assessments to ethics boards auditing enterprise risk across pillars of trust. Dynamic regulations that address issues such as privacy rights and encryption by design principles prevent intrusive data collection without deterring lawful functionality. Platform governance reforms can curb addictive business models and amplify pollutants. Updating fiduciary duties to balance shareholder returns and multi-stakeholder impacts steers corporate capacity responsibly. Ultimately, decision-makers should evaluate AI applications based on their contribution to the holistic flourishing of lives rather than just narrowly defined task excellence. As exponential technologies continue to accelerate, conscientious development must ensure that AI's benefits empower society inclusively. In addition to supporting rapid innovation, governance plays a crucial role in guaranteeing accessible infrastructure, closing divides, and directing capabilities justly to uplift marginalized groups [286]. Progress measures should not only aggregate advances but also track the closing of gaps by gender, race, region, and income to ensure that advances reach and aid those who may be overlooked by default. Policy mechanisms that promote equitable access span funding to close broadband gaps that inhibit rural business opportunities enabling microcredit supplied by responsibly regulated alternative lenders and allowing public sector data use for targeting programs to communities historically denied services. Precedents such as open-access scientific publishing, public domain media, and universal service programs provide models that prevent monopolies from inhibiting affordability and legal access. With vigilance, AI can democratize prosperity rather than concentrate its dividends in the hands of a select few. Artificial intelligence holds immense potential to uplift global living standards, provided that governance steers progress towards addressing human needs and higher ideals instead of focusing solely on efficiency. Today's leaders bear the responsibility of implementing AI applications thoughtfully as a directional choice rather than a deterministic tide. Policies that promote work liberating human creativity, direct algorithms towards societal priorities, ensure representative development and oversight, take full social costs into account, and guarantee broad affordability promise to raise wisdom over accumulating undirected power. Conscientious collaboration can help ensure that technological transformation serves the empowerment of all.

4.3 Embracing Change & Envisioning the Future - Strategies for Individuals and Organisations & Positive Vision of AI-driven Workplace

4.3.1 Adapting to the Future of Work as Individuals

The swift progress of technologies like artificial intelligence and automation has brought about a transformation in job markets, necessitating individuals to adapt to changing skill requirements and novel hybrid human-AI augmented workflows. In addition to enhancing their technical expertise, individuals ought to concentrate on cultivating a mindset of lifelong learning, creativity, networking, inspiration-seeking, and entrepreneurialism to sustain their competitiveness during unstable times. Furthermore, maintaining a positive outlook motivation and perceiving technology as an empowering force rather than a threatening one can facilitate personal transitions. Individuals can effectively navigate the workplace transformation wave by upgrading their perspectives and abilities and capitalising on new opportunities.

Developing a growth mentality that centres on lifelong learning and consistently upgrading one's skills is imperative for individual adaptability and resilience amid technological change. According to research, perceiving self-efficacy and considering skills as continually developable through effort is predictive of adopting a positive attitude toward new challenges, which fixed mindsets interpret as threats to their inability [305]. Since technological skills constantly evolve, avoiding obsolescence necessitates regular learning of emerging tools, capabilities, and paradigm shifts while deeply applying established domains. This requires individuals to proactively seek educational opportunities, such as online programs, professional conferences, mentorships, and lateral job rotations that expose them to new specialities. For instance, an accountant may opt for a crash course on Robotic Process Automation (RPA) software that automates financial tasks while enhancing interpersonal advisory skills. Being mindful of avoiding cognitive biases and outdated assumptions further aids in continually refreshing perspectives. Overall, the growth mentality's learning-apply-repeat cycle promises sustained employability as work environments transform.

While human strengths like creativity continue to be crucial, all individuals must possess foundational digital literacy and be comfortable adopting new technologies to contribute amidst the integration of AI and automation trends. Global surveys reveal that despite acknowledging the rising impact of technology in the workplace, many professionals acknowledge skill gaps around data analysis, information validation, software adoption, and even basic computer troubleshooting that impede their effective utilisation of emerging tools [306]. Addressing such skill gaps requires an honest self-assessment of competencies across domains such as basic coding, analytics, privacy ethics, and prototyping mobile applications. Free online learning platforms now enable accessing micro-courses targeted at upskilling. Regularly using shared collaboration suites for non-digital roles improves interfacing with data-driven decision-making and project management. Individuals must proactively allocate time routinely to upgrade their tech skills and not presume that one-off training suffices. Workplace automation necessitates a persistently curious approach toward developing technological literacy and digital capabilities complementary to human specialities. As artificial intelligence increasingly matches or exceeds human capabilities on analytical and repetitive tasks, the significance of unique human strengths, such as imagination, creativity, emotional intelligence, and ethics, becomes more central to augmented job designs [288]. Individual workers must identify, develop, and allocate more time toward such irreplaceable skills that machines lack as automation complements technical and routine contributions. For instance, disciplines such as design, writing, strategic foresight, and anthropological inquiry leverage fundamental human gifts around pattern recognition, empathy, communication, and abstraction. Enterprises will increasingly turn to creative positions as automation handles scales, such as graphic designers crafting visionary concepts informed by data patterns. Therefore, individuals should devote time to exploring and honing creative pursuits they find intellectually engaging and socially meaningful, regardless of prior professional paths. Lifelong learning includes actualising talents that automation makes currently relevant across various emerging roles.

In times of uncertainty, sharing valuable information, mentoring, and awareness of new opportunities often occur through trusted professional networks and informal social connections. Therefore, it is crucial for individuals to continually build bridges and participate in both in-person and digital communities to gain access to insider insights, trusted guidance, and potential collaborators or sponsors that upskilling in isolation may overlook [307]. This requires regular engagement in events, conferences, and online forums while actively maintaining existing contacts by staying in touch and sharing useful content. Networking also involves politely seeking short informational interviews from experienced experts one wishes to emulate, sometimes requiring calculated persistence. Over time, by thoughtfully nurturing genuine mutual relationships, individuals gain sounding boards providing honest feedback, partners in solving emerging problems, and sponsors advocating for their readiness amid new openings. When combined with skill development, reliable, professional networks enable individuals to seize opportunities confidently in evolving work environments. Finding mentors with hard-earned experience in navigating workplace change during times of disruption and paradigm shifts can provide invaluable counsel and inspiration for individuals. Mentorship relationships can guide individuals in setting realistic development goals, focusing on self-improvement energy, and modelling positive mindsets that embrace uncertainty as an opportunity [308]. Great mentors can decode unwritten cultural rules, share tacit high-impact work habits, and demonstrate that massive transformation is navigable with thoughtful strategy. In practice, individuals should humbly seek advice from respected figures by conducting informational interviews about their journey. Conferences, trade associations, and alumni groups can provide access to these figures. Once an organic connection is established, protégés can seek guidance at pivotal moments while being careful not to impose too frequently out of genuine respect. Over time, authentic mentor bonds can crystallise through continually earning trust and paying it forward. Individuals can stay encouraged and strategically adaptive by learning from those who have overcome past disruptions. The fear of workplace automation may lead individuals to become anxious and cynical, resulting in self-defeating pessimism and disengagement. To counter this, it is suggested that individuals intentionally cultivate reality-based optimism, recognising that automation is a part of progress's perpetual cycle. While the transition may bring unease, humans have historically shepherded technological advances that have benefited society while creatively advancing their roles [309]. Individuals can spur proactivity and shape their motivational mindset by reframing progress as an opportunity.

It is important to consciously limit the consumption of scaremongering misinformation and doomsday hyperbole and, instead, ground oneself in balanced perspectives through solutions-focused educational content. Negativity often breeds complacency rather than constructive action. Shared struggles can unite colleagues to swap transition tips, and gratitude, social support, and self-care can help manage stress and prevent burnout. With disciplined, positive mindsets, individuals can stay energised and creatively improve their circumstances rather than being overwhelmed by forces beyond their control. In addition, adapting to the future of work requires individuals to cultivate their latent entrepreneurial spirit to identify problems that require solutions and imagine possibilities amidst uncertainty. According to research, entrepreneurs often envision fresh opportunities and worthwhile ventures by pursuing inherent interests, whereas organisation-centric employees await detailed instructions rather than ideating options [310]. Individuals should not view automation's business and job churn as a threat but rather as an opportunity to create new roles, specialities, and ventures. Through creativity and calculated risk-taking, they can explore adjacencies, niches, and emerging capabilities to spark commercial or social enterprise concepts, collaborators, and potential backers. By doing so, individuals can identify persistent problems that lack current solutions and devise innovative approaches to address them. The sheer pace of workplace transformation necessitates enterprising individuals who can identify needs and boldly organise resources to test solutions. In this way, they can remain indispensable even as machines handle old business models. Therefore, individuals should embrace their entrepreneurial spirit and engage in creative ideation to identify new opportunities and ventures that will enable them to thrive in the changing world of work. The advent of workforce automation technologies necessitates that individuals prepare by developing new skills and mindsets to remain professionally competitive and unlock new possibilities. In addition to improving technical literacy, individuals should pursue lifelong learning, engage in creative endeavours, cultivate trusted relationships, seek out inspirational mentors, and explore entrepreneurial opportunities that reveal their full potential while machines transform once-static work environments. By upgrading their human capital and adaptive abilities, individuals can confidently embrace workplace change as progress that enables new prosperity rather than a threat. They can develop the skills and mindset needed to thrive in a rapidly changing work environment and unlock new opportunities that were previously unavailable. Therefore, individuals need to be proactive in their professional development and seek opportunities for personal and career growth. Through continuous learning and creative pursuits, individuals can develop the skills and knowledge needed to stay competitive in the face of technological disruption. By cultivating trusted relationships and seeking inspirational mentors, they can gain valuable insights and guidance to navigate the changing work landscape. Through entrepreneurial opportunities, individuals can unlock new possibilities and realise their full potential, even as machines transform the nature of work.

4.3.2 Opportunities for Business Leaders

The incorporation of artificial intelligence (AI) in the work environment presents a significant prospect for corporate leaders to express a motivational perspective for the future of their establishments. With the transformation of work processes through automation and AI, leaders must establish a connection between technological advancements and elevated purpose. An engaging vision can foster a sense of shared aim among employees, inspiring them to accept change [311]. To develop an inspiring vision, leaders must first determine the fundamental values and aspirations that will steer the organization towards the future. For instance, a leader may articulate a vision of a future where technology enables the enterprise to provide innovative services to customers. They can link this vision to values such as innovation, customer service, and meaningful work for employees. By specifying the benefits, the vision becomes more tangible and captivating. Furthermore, leaders should recognise potential apprehensions associated with AI. A vision rooted in shared values can reassure individuals that new technologies will be implemented with ethical considerations. For instance, a leader can commit to providing opportunities for employees to learn new skills as job roles evolve. Presenting AI as a means of eliminating mundane tasks and enhancing human capabilities can alleviate concerns [312]. To actualise the vision, leaders must convey it regularly through various channels. Corporate events, emails, and informal discussions serve as points of contact to reinforce the narrative consistently. When individuals hear the vision linked to their daily routines, it becomes more palpable and motivating [313]. A lucid vision driven by values can unite and inspire individuals to navigate the future confidently in the era of technology.

With AI revolutionising job roles across various industries, continuous learning will become essential for employees to develop new skills. Business leaders have a crucial role to play in providing resources and incentives to facilitate retraining initiatives in the AI-driven workplace. Firstly, leaders must allocate resources towards training programs, partnerships, and platforms to enable continuous skill development. Online courses, periodic skill-building workshops, and tuition reimbursement policies are examples of mechanisms that leaders can establish [314]. In addition, partnerships with online education providers, coding bootcamps, and academic institutions expand employee learning access. Furthermore, leaders should consider incentives that motivate employees to engage in retraining efforts. Tuition reimbursement, bonuses for credentials and certifications, and increased compensation for acquiring new skills are all ways to encourage learning. Public recognition of accomplishments through company awards or highlighted profiles also helps incentivise retraining. Leaders need to establish a culture that prioritises continuous skill enhancement. They can promote learning opportunities through company communications and events. Leaders can signal the importance of skill development by leading by example and undertaking retraining themselves. Providing daily or weekly time allowances for learning also creates an atmosphere where skill-building becomes a workplace norm. By providing sufficient resources and incentives for retraining in an AI environment, leaders empower their workforce to adapt continuously [314]. This benefits both employees by keeping their skills current and marketable and the organisation by building a competent, future-ready team. The AI shift necessitates leaders to take an active role in facilitating ongoing staff learning. The integration of AI into organisations has the potential to introduce new products, services, and efficiencies. However, this can only be achieved with a culture that welcomes innovation and experimentation. As AI continues to transform the workplace, business leaders play a crucial role in fostering conditions that encourage innovation. Leaders must provide employees with the necessary resources, incentives, and permission to experiment and explore new ideas. This can be achieved through various means, such as hackathons, innovation labs, and designated days for experimentation. Allocating funding for pilot projects and prototypes also provides resources for innovation. Furthermore, recognition and rewards for experimentation efforts, even those that are unsuccessful, reinforce the value of innovative risk-taking. Encouraging collaboration among teams and functions is another key factor in promoting innovation. Business leaders can create cross-functional groups and platforms that facilitate brainstorming ideas across silos. Additionally, relaxed spaces that are designed for ad-hoc collaboration can encourage natural innovation "collisions" to occur [315]. It is important for leaders to establish psychological safety, which allows individuals to take risks without fear of failure [316]. Leaders can demonstrate this by admitting their missteps and showing that mistakes are acceptable. Reframing failures as learning opportunities can also make experimentation less daunting. With psychological safety, individuals feel empowered to turn their ideas into transformative innovations. Embracing experimentation is a critical step for leaders to enable their workforce to leverage AI as a springboard for innovation. Employees who continuously build, test, and refine new solutions can position the organization to lead amidst AI-driven disruption. While AI systems generate valuable insights and recommendations that can inform better decisions in the workplace, centralizing all decision-making authority with AI or management often disempowers employees closest to work. Business leaders have an opportunity to pair AI with more decentralized, democratic decision-making processes to fully engage their workforce. Empowering frontline teams with autonomy and authority over local decisions can unleash their contextual expertise [317]. For instance, it enhances responsiveness by granting customer support teams the flexibility to resolve complaints using their judgment without manager approval. Leaders should provide access to AI insights as input rather than a mandate for devolved decision-making teams. For instance, sharing demand forecast data can allow production teams to weigh this intelligence against other factors when determining how to efficiently allocate resources [318]. Decentralized decision-making teams should represent diverse internal and external perspectives. Inviting end-users or outside experts to contribute insights can make decisions more robust and transparent [319]. AI can also help identify gaps in decision-making groups to promote diversity of thought. Empowering employees sustains engagement in an era of AI, which threatens to marginalise human expertise. Combining AI insights with decentralised, inclusive decision-making teams leverages technology's potential while keeping humans in the loop.

The socio-technical challenges of governing AI and other emerging technologies are too complex for any leader or organisation to solve alone. As a result, business leaders should look beyond their walls and collaborate across sectors to access wider expertise for the AI-integrated workplace. Partnering with civil society groups, academics, and public agencies can expand understanding of AI. For instance, collaborating with researchers to study algorithmic bias can help leaders create more ethical systems [320]. Expert panels with diverse ideologies can surface innovative ideas and identify potential harms [273]. Cross-sector partnerships can also pilot solutions to complex challenges. Companies may collaborate with governments and nonprofits to test AI applications that serve public interests, such as healthcare access or sustainable development [294]. Policy labs that bring together public, private, and civic leaders can enable proactive governance approaches to be co-designed rather than reactively regulated. Collaborative policy development can benefit issues such as responsible data stewardship, AI transparency, and mitigating workforce disruption [321]. Tackling the risks of AI and harnessing its potential requires new forms of collective leadership and responsibility across stakeholder groups. By proactively seeking diverse expertise through cross-sector collaboration, business leaders can make more informed and ethical decisions when deploying AI. Although AI holds enormous promise, leaders are responsible for ensuring that these technologies are deployed ethically and for broad human benefit. Business leaders must make AI ethics, fairness, and human welfare core priorities if they hope to sustain responsible innovation. Ethics governance mechanisms, such as review boards and AI codes of conduct, institutionalize consideration of societal impact [322]. Human rights impact assessments can uncover potential issues such as privacy infringement or disproportionate effects on marginalised groups that undermine human welfare [323]. Risk monitoring procedures help organisations continuously evaluate and address emerging harms.

Leaders should prioritize diversity and interdisciplinary collaboration in AI teams to minimise harmful bias. Psychologists, anthropologists, and subject matter experts can identify risks that computer scientists often overlook [324]. Workforce training in ethical thinking builds a collective capacity to recognise concerns. Most importantly, leaders must shift mindsets and incentive structures so that ethics and human values become central considerations, not afterthoughts. Evaluation criteria should emphasise social responsibility alongside technical accuracy or business returns [325]. A shared vision rooted in human dignity and flourishing makes ethics intrinsic to organisational culture. Establishing partnerships and two-way exchanges with technical experts, academics, and innovators generates valuable intelligence. Advisory boards, visiting tech fellows programs, and university collaborations facilitate knowledge sharing on emerging developments [326]. Site visits to pioneer facilities engaged in leading research also provide firsthand learning opportunities. Continuously learning about unfolding possibilities empowers leaders to seize opportunities enabled by AI more confidently while wisely navigating risks. Investing to understand ongoing technology advances yields the insights required for robust strategic planning.

4.3.3 Strategies for Policymakers

Rewrite academically. Pay attention to clarity. Do not separate into paragraphs. The transformative potential of artificial intelligence relies upon universal, affordable access to high-speed internet and connected devices. Policymakers play a vital role in funding digital infrastructure and inclusion programs to prevent groups and regions from being left behind in an AI-driven workplace. Major investments in broadband networks, especially fibre optics, 5G and satellite internet, provide the backbone for data transmission and next-generation technologies [327]. Grants, loans and public-private partnerships can fund infrastructure in rural and low-income areas with less commercial incentive. Sponsoring municipal broadband also increases affordable access. In order to ensure the inclusive adoption of AI, policymakers should provide funding and frameworks for universal and participatory access to emerging technologies. Device access initiatives such as long-term device financing, public computer labs, and digital literacy training should be funded to achieve this. Voucher programs can also be implemented to help students, older adults, and low-income households participate fully in the AI economy. In addition, workforce policies must ensure internet affordability and utility-style regulation can be introduced to curb harmful monopolies. Subsidies can also be utilised to make home broadband more accessible to vulnerable groups. For instance, internet costs can be included in existing housing assistance programs. By closing digital divides, policymakers can lay the groundwork for an inclusive AI adoption.

In order to fully realize the potential of AI, policymakers should promote "computational literacy" at all levels of education. This involves equipping students with the necessary concepts and skills for the AI-driven workplace. At the K-12 level, computer science and data literacy should be core requirements instead of electives. To achieve this, funding should be provided for teacher training, curriculum development, and classroom technology. This will enable all students to achieve computational fluency before college or entering the workforce [328]. In addition, higher education policies should incentivise degrees that combine technical and non-technical disciplines. As AI continues to influence roles across various sectors, having a technical grounding coupled with application knowledge can create versatile and collaborative workers. Apprenticeships and vocational programs can also integrate classroom and on-the-job technology skill-building, allowing students to gain practical experience while developing their technical skills.

To promote lifelong computational literacy, policymakers can offer tax incentives for continuing education and professional training in digital capabilities, encouraging individuals to continuously update their skills and stay relevant in an AI-driven workplace [299]. In addition, public digital skill platforms and community-wide workshops can be established to promote inclusion and ensure that everyone has access to the necessary resources. By securing knowledge foundations through education policy, citizens can actively shape the unfolding of AI rather than passively experience technologies handed down to them. This will ensure that individuals are equipped with the necessary skills and knowledge to fully participate in and benefit from the AI economy.

Geographic hubs that concentrate talent, funding, and infrastructure in specific technology sectors have proven to be powerful engines of innovation. To position their regions or nations competitively, policymakers should develop incentives and partnerships to foster the formation of AI research clusters. This can be achieved through tax credits, grants, and specialized facilities to attract anchor companies to locate research centres, which will then stimulate local ecosystem growth. Drawing academic programs and startups builds a skilled talent pool, while business parks and incubators provide shared spaces for collaboration across entities. Policymakers can foster innovation and growth by creating AI research clusters, leading to economic and societal benefits.

Sponsoring applied research at local universities can be an effective way to orient projects toward regional strengths. This approach ensures that research is relevant to the local context and can be applied to address regional challenges. Furthermore, cluster connectivity can be facilitated through convening events, digital platforms, and research commercialization programs [329]. At the national level, funding agencies can play a critical role in fostering innovation and growth by coordinating countrywide initiatives like Grand Challenges. Such initiatives coalesce research around large societal goals, leading to a more coordinated and effective approach to addressing national challenges. [330]. In addition, defence research investments have been found to further catalyze mutually reinforcing innovation. This approach can lead to advancements in both military and civilian technology. By leveraging these varied strategies, policymakers can foster innovation and growth on both regional and national levels. Nurturing AI clusters can create positive feedback loops, with each new contributor expanding the realm of possibilities. Policymakers can activate this process by establishing optimal conditions for density, diversity, and connectivity. However, while technological disruption creates new opportunities, it also threatens to displace roles, worsen inequality, and erode economic security. Policymakers have a duty to update worker protections and strengthen social safety nets as AI transforms the labour market. To achieve this, policies should safeguard fair wages, prevent misclassification of gig workers, and guarantee basic rights like family leave regardless of employment category, as Acemoglu and Restrepo [262] suggested. By implementing these policies, policymakers can ensure that the benefits of AI are equitably distributed and that individuals are protected from potential negative impacts.

Assisting workers during transitions is crucial to ensuring the well-being of individuals in the workforce. Wage insurance is one strategy that can be employed to temporarily replace lost earnings in the aftermath of displacement. Other measures, such as mobility grants, career navigators, and skills training, can help people transition into new roles over the long term. Relocation vouchers can also facilitate the pursuit of opportunities in technology hubs [331]. To support workers further, robust social safety nets can be implemented to decouple basic economic security and healthcare from specific employers, enabling more fluid transitions [332]. Paid leave, child allowances, and subsidised relocation can also build individual resiliency in the face of disruption. Additionally, automation taxes could be implemented to fund transitional supports and lifelong learning accounts [333]. It is important to note that thoughtful policies can ensure that AI promotes inclusive prosperity rather than social bifurcation.

The rapidly evolving nature of AI innovation creates a risk of outpacing governance capacity, requiring policymakers to play a crucial role in convening diverse stakeholders to develop shared standards and decision frameworks that steer technology for the greater good. Lifecycle analysis standards assist developers in assessing social impacts at each phase, from design requirements to testing protocols and monitoring frameworks post-deployment [334]. Regulating explainability and transparency improves interpretability for regulators and the public [335]. Sector-specific frameworks allow tailored oversight appropriate to different risks. For instance, healthcare AI necessitates stringent validation to uphold safety and privacy [336]. Rules that ease access to training datasets promote innovation while securing data rights. Inclusive decision-making bodies can dynamically develop policies. For example, AI ethics boards with diverse ideologies debate tradeoffs between principles such as safety, privacy, and access [334]. International collaboration also aids governance, given the borderless data flows [337]. Standards can evolve responsibly alongside technologies by establishing flexible, context-specific frameworks through collective deliberation. Policymakers play a crucial role in structuring democratic processes for AI governance. AI governance is a domain that is fraught with uncertainties. Policymakers should launch pilot programs that test different models to generate evidence and experience for designing long-term policies. Sandbox experiments in defined geographic areas or industries provide a contained environment to assess policy options. For example, exploring how a combination of AI workforce tax and universal basic income impacts economic security and opportunity [338]. Public-private partnerships enable studying policies at scale. Partnerships that expand vocational AI training or ban certain uses in law enforcement provide real-world learning [339]. Policy labs can also simulate alternative models to identify risks beforehand [321]. By conducting these experiments, policymakers can gain a better understanding of the potential implications of different policies and use the evidence generated to make informed decisions on long-term policy design.

Forward-looking policies, such as universal broadband, life-long learning accounts, and flexible labour laws, empower society to adapt to future shifts. Periodic review processes allow policies to evolve responsively while providing stability for long-range planning [340]. Fiscal policies should factor in the cumulative effects of automation rather than yearly budget cycles [341]. Models projecting economic, governmental, and societal effects over 20-50-year horizons reveal policy gaps. Convening diverse foresight exercises creates a collective vision to shape policy. Scenario planning surfaces key variables and inflexion points through multi-stakeholder dialogue [342]. Speculative prototyping makes societal futures tangible [343]. The lasting governance frameworks needed for socially beneficial AI will not be designed overnight but will gradually emerge through sustained learning, values-driven vision, and proactive adaptation. Policymakers play a crucial role in this long-game approach. By promoting forward-looking policies and fostering a collaborative environment for foresight exercises, policymakers can help society adapt to future shifts and create lasting governance frameworks supporting socially beneficial AI development.

4.3.4   *Positive Vision for Human-AI Collaboration*

Machines would handle repetitive analytical tasks in an ideal future powered by human-AI collaboration, enabling more people to pursue creative careers aligned with their passions. AI systems excel at pattern recognition, prediction, and optimization, while humans spark breakthroughs through imagination, empathy, and an appreciation of beauty [344]. Delegating rote jobs to AI frees humans to dedicate their energy toward original artistic expression, design thinking, and cultural innovation. By leveraging the strengths of both humans and AI, society can create a future where people are empowered to pursue fulfilling careers while machines handle tasks that are better suited to their capabilities. This vision of the future highlights a world where creative expression and innovation are given a higher priority, leading to a more enriched society. With basic needs reliably met through AI automation, the economics of creative careers become more viable for the masses. Generative AI also assists the artistic process by providing inspiration on demand or augmenting human creativity. For example, an architect may leverage AI to rapidly iterate design variations while concentrating their unique vision on the feeling the space should evoke. This future celebrates human ingenuity and our innate need for self-expression. With AI handling routine analytical tasks, people across all backgrounds and aptitudes can spend their time bringing creative ideas to life through art, literature, design, music, and innovation. Ubiquitous creativity services raise the quality of life while also driving economic dynamism as new cultural experiences stimulate further invention. In the future, individuals can pursue fulfilling careers in creative fields that were previously inaccessible or impractical. By leveraging the capabilities of AI, society can create an environment where human creativity is unleashed on a wider scale, leading to a more enriched and diverse cultural landscape.

In an ideal future of human-AI collaboration, meaningful work would provide dignity and purpose for all people regardless of occupation. As many routine jobs become automated, new roles would emerge coordinating and directing AI systems toward solving complex challenges that benefit society. Work would become focused on human skills like emotional intelligence, creativity, cross-cultural collaboration, and ethical reasoning that technology cannot replicate. Rather than income, the purpose would become the driving motivation for work [338]. Regular hours would decline as task-oriented roles fluidly form around tackling specific challenges. Policies like universal basic income guarantee economic security decoupled from specific jobs, providing freedom to pursue meaningful projects [345]. In this future, individuals would have the opportunity to pursue work that aligns with their passions, values, and strengths, leading to a more fulfilled and purpose-driven society. By leveraging AI to automate routine tasks, society can create a future where work is focused on the uniquely human abilities that add value to society. In the context of organizational management, leaders play a fundamental role in shaping company cultures that prioritize the well-being of individuals over short-term financial gains. In order to ensure ethical deployment, AI systems require continuous oversight. Empowered teams should retain decision-making power, guided by organizational values. The role of technology should be to enhance human flourishing rather than to replace people. This vision of the future is one in which AI is leveraged to elevate the aspects of work that are meaningful and dignifying while eliminating dangerous and dehumanizing jobs. As a result, individuals experience a sense of purpose and value in their work, aligning with a higher organizational purpose. In a hypothetical future scenario, artificial intelligence (AI) would be leveraged to optimize complex societal systems, ensuring efficient provision of fundamental necessities such as healthcare, education, housing, and infrastructure. By coordinating data and decision-making across fragmented systems, machine learning has the potential to address challenges such as healthcare, environmental sustainability, and social inequality that have eluded human governance [238]. For instance, AI systems can enhance healthcare delivery by predicting patient needs, tracking resources, and coordinating care providers, thereby reducing costs and improving patient outcomes [346]. Moreover, intelligent infrastructure can manage transportation patterns, energy grids, and waste systems, making cities more livable and sustainable. Finally, supply chains can be optimized to balance production, inventory, and logistics to prevent shortages and overproduction.

By freeing human experts from mundane tasks, AI enables them to focus on high-level strategy and oversight. Additionally, ecosystem-level insights provided by AI systems inspire humans to devise innovative and systemic solutions that contribute to the betterment of society. As basic foundational systems run smoothly with the help of AI, greater resources become available to fund human services such as education and the arts. Overall, AI's system optimization capabilities have the potential to elevate the well-being of society as a whole. In an ideal future, governance frameworks would be in place to ensure that AI-based decisions at all levels are informed by ethical considerations and reflect diverse human perspectives. Open-sourced algorithms would allow broad communities to continuously debug biases. Furthermore, participatory design practices would empower impacted groups to shape technology that addresses their specific needs [347]. To guide context-specific policy development, global AI ethics boards comprising diverse ideologues, spiritual leaders, and philosophers would constantly debate principles and values [337]. Additionally, rigorous audits would assess algorithms' real-world impacts before deployment. Ongoing monitoring would also be a key feature of such governance frameworks. In the envisioned future, decision-making authority is not entirely delegated to AI. Instead, human insights play a crucial role in contextualizing machine recommendations. For instance, doctors and nurses interpret medical AI in light of a patient's preferences and psychosocial needs. Furthermore, democratic, multidisciplinary teams participate in key organizational decisions aided by AI insights. This future embraces AI but ensures that humans remain firmly involved in ethical governance. Algorithms are designed to enhance rather than replace human judgment and moral wisdom.

In the envisioned future, social contracts are redefined to ensure that economic prosperity and rewarding work are broadly shared, particularly in light of the transformative impact of AI on both society and the economy. Universal programs guarantee fundamental needs such as healthcare, education, and child allowances as human rights, thereby providing flexibility amidst technological shifts. Policies such as progressive taxation on automated work, data dividends, and stakeholder shareholding are implemented to spread capital ownership and the value created by AI across society [338]. Additionally, profit-sharing schemes are implemented to give workers bargaining power to demand fair compensation as technology increases productivity. In the ideal future, labour laws are designed to reinforce human dignity and prevent exploitation. Standardized policies such as minimum living wages, responsive work schedules, parental leave, and caregiving support are applicable to all workers, including independent contractors in the gig economy. Mobility grants are also provided to support worker transitions and geographic relocation for new opportunities. This inclusive social contract empowers people to participate fully as technology transforms society. AI is designed to increase prosperity for many rather than wealth for a select few. Entrepreneurship and grassroots innovation flourish in the envisioned future as access to knowledge resources, funding, and digital infrastructure becomes ubiquitous. Online education, mentorship platforms, and open-sourced research enable anyone to gain expertise for turning ideas into reality [348]. Generative AI assists startups with planning, design, and business operations. Furthermore, decentralised autonomous organisations allow collectives to coordinate passion projects using collaborative tools [349]. Regulators play a crucial role in balancing open data access with personal rights to prevent surveillance while fostering innovation.

In the envisioned future, policies such as basic income, flexible employment laws, and convertible visas are implemented to provide individuals with the freedom and security to experiment and take risks without jeopardising their livelihoods. State-funded innovation clusters and business accelerators are also established to nurture local ecosystems around shared strengths and culture. This entrepreneurial future empowers broad segments of society to invent, build community wealth, and pursue purpose-driven careers. AI has the potential to unlock a paradigm of mass innovation. In the envisioned future, lifelong education is designed to cultivate not only technical skills but also creativity, emotional intelligence, ethics, cultural knowledge, and self-understanding. The goal shifts from maximizing wealth to maximizing human development and living in a manner aligned with one's unique identity and purpose [350]. Standardised curriculums are replaced by personalised, project-based learning tailored to diverse abilities and interests. Course offerings span everything from wilderness survival to the anthropology of storytelling. Apprenticeships provide contextual skill-building in real-world environments. Additionally, accessible virtual worlds enable safe spaces to experiment with new identities and experiences. In the envisioned future, teachers assume the role of mentors, guiding independent exploration across domains ranging from craft to sport to spiritual wisdom. Policies such as tuition subsidies, income-sharing, and career guidance are put in place to fund nontraditional educational pathways. Furthermore, lifelong upskilling and sabbaticals are made available to prevent stagnation. In this future, education moves beyond work-readiness, focusing instead on nurturing well-rounded lives. Learning equips individuals to flourish amidst change, steer technology ethically, and find meaning by actualising one's highest self.

Chapter 5

Introduction to the Future of Work with AI

5.1 Historical Evolution of Work and Technology

The interplay of work and technology, tracing its roots from ancient civilizations to the modern era, has been pivotal in shaping human societies. Each technological advancement has redefined the nature of work, leading to profound societal transformations.

Early Civilizations and the Birth of Structured Labor.

The inception of structured labor can be traced back to ancient civilizations where agriculture and the establishment of cities necessitated a more organized approach to work. This era saw the emergence of diverse professions and the early division of labor, as documented by historians like Diamond (1997) in his exploration of human societies [351].

The Industrial Revolution: A Paradigm Shift.

Marking a significant departure from agrarian societies, the Industrial Revolution introduced mechanization and changed the landscape of work dramatically. The rise of factories and urbanization, as discussed by Landes (1969), brought about a shift from manual craftsmanship to mass production, reshaping social structures and the concept of employment [352].

The Information Age and Digital Transformation.

With the advent of computers and the Internet, the late 20th century heralded the Information Age. This period, characterized by rapid technological advancements in digital communication and information processing, as Bell (1973) describes, revolutionized the workplace, making information and communication technology an integral part of professional and personal life [353].

Rise of AI and Its Impact on Work.

The emergence of artificial intelligence (AI) in the 21st century represents the latest evolution in the symbiosis of work and technology. AI's potential in automating complex tasks, as elaborated by Kaplan and Haenlein (2019), is not only changing the nature of work but is also raising questions about the future of employment, ethical considerations, and the balance between work and leisure [354].

Technological Evolution and the Concept of Free Time.

The article "Creating the Super Free-Time Society" by Professor Ichiya Nakamura[1] posits an intriguing vision of the future, where advances in AI and robotics may lead to a society with significantly more leisure time. This notion, while optimistic, requires careful consideration of the economic, social, and ethical implications of such a transformative shift, as discussed by Schwab (2017) in his analysis of the Fourth Industrial Revolution [309].

5.1.1 Defining the AI-Driven Workplace

The Advent of the AI-Driven Workplace.

In today's rapidly evolving work landscape, the AI-driven workplace is becoming increasingly prevalent. This concept extends beyond the mere implementation of technology; it represents a fundamental shift in the way work is conceptualized and executed. The integration of AI technologies, such as machine learning, natural language processing, and advanced analytics, is reshaping the dynamics of various industries. Lee (2018) provides insight into this transformation, highlighting the interplay between human intelligence and AI's computational power in shaping the future of work [22].

Characteristics of the AI-Driven Workplace.

The AI-driven workplace is marked by several defining characteristics. First, there is an emphasis on efficiency and productivity, with AI systems automating routine and repetitive tasks. This shift allows human workers to focus on more creative and strategic tasks. Second, AI enables data-driven decision-making,

---

[1]https://ichiyanakamura-en.blogspot.com/2023/04/new-edition-of-creating-super-free.html.

where vast amounts of data can be analyzed to inform business strategies and operations. Bostrom (2014) discusses the potential of AI to transform decision-making processes in various sectors [355]. Third, AI technologies offer personalized experiences, tailoring services and products to individual needs and preferences.

Impact on Workforce Dynamics.

The integration of AI into the workplace is significantly altering workforce dynamics. On one hand, there are concerns about job displacement due to automation. On the other hand, AI is creating new job opportunities in fields related to AI development, management, and ethical governance. Manyika et al. (2017) explore these dual aspects of AI in the labor market, emphasizing the need for policy and educational responses to manage the transition [356]. Additionally, the workforce is increasingly required to adapt to AI systems, necessitating continuous learning and skill development.

Ethical and Social Considerations.

The proliferation of AI in the workplace brings with it a host of ethical and social considerations. Issues of privacy, data security, and the potential for bias in AI algorithms are of paramount concern. Wallach and Allen (2009) delve into the moral dimensions of AI, arguing for the importance of aligning AI development with ethical principles and societal values [357]. Furthermore, the potential for a digital divide, where access to AI technologies is unequally distributed, raises questions about equity and inclusion in the AI-driven future.

The Future Outlook of AI in Work.

As we look towards the future, the role of AI in the workplace is set to deepen and expand. The integration of AI with other emerging technologies such as blockchain, the Internet of Things (IoT), and advanced robotics will likely lead to further innovations and transformations. Schwab (2017) captures this vision in his discussion of the Fourth Industrial Revolution, where technological advancements are not only reshaping industries but also the very fabric of society [309].

5.1.2 *Impact on Professional and Personal Interactions*

Revolutionizing Professional Communication.

The integration of AI in professional communication has led to significant improvements in efficiency and effectiveness. In the corporate world, AI-powered tools are being used for customer service, data analysis, and decision-making processes. This shift is not only enhancing the speed and accuracy of communication but is also enabling more personalized customer experiences. A study by Homburg et al. (2015) highlights how AI-driven customer interaction systems can improve engagement and satisfaction [358].

Transforming Personal Communication.

On a personal level, AI has revolutionized the way individuals interact. Social media platforms, equipped with AI algorithms, are curating personalized content, thereby shaping how information is received and perceived. This personalization, as Turkle (2011) observes, is altering the nature of social interactions and relationships in the digital age [359]. Additionally, AI-driven translation services are bridging language barriers, facilitating global communication like never before.

AI in Interpersonal Skills and Empathy.

One significant aspect of AI in communication is its potential to understand and simulate human emotions. AI systems, like emotional chatbots, are being developed to recognize and respond to human emotions, aiming to enhance empathy in digital interactions. Picard (1997) in her seminal work on affective computing discusses the prospects and challenges of imbuing AI with emotional intelligence [360].

Challenges in AI-Mediated Communication.

Despite the advancements, AI-mediated communication faces several challenges. Issues of privacy, data security, and the potential for miscommunication remain concerns. Furthermore, the reliance on AI for communication can lead to a reduction in face-to-face interactions, potentially impacting social skills and relationship building. Goffman (1959) in his analysis of interpersonal communication underscores the importance of direct human interactions [361].

The Future of AI in Communication.

Looking ahead, the role of AI in shaping both professional and personal communication is poised to grow. As AI technologies become more sophisticated, their ability to facilitate and enhance human interaction will likely increase, albeit with ongoing challenges and ethical considerations. The work of Floridi (2014) on the ethics of information provides a framework for understanding these future developments [362].

5.2 Digitization of Media and Information Dissemination

5.2.1 AI in News and Entertainment Industries

AI's Role in the News Industry.

The news industry has undergone a significant transformation with the advent of AI. Artificial intelligence is reshaping how news is gathered, reported, and distributed. Algorithms are now used to sift through vast amounts of data to identify newsworthy trends and even generate news reports in some instances. Napoli (2019) discusses the implications of AI in the media landscape, highlighting both the potential for enhanced news dissemination and the challenges related to journalistic integrity and algorithmic bias [363].

Personalization and Targeting in News.

One of the most notable impacts of AI in the news industry is the ability to personalize content for viewers and readers. AI systems analyze user preferences and behavior to deliver tailored news feeds, thereby significantly altering the traditional one-size-fits-all model of news dissemination. Thurman and Schifferes (2012) explore the growing trend of personalized news, examining its effects on the diversity of information consumption [364].

Revolutionizing the Entertainment Industry.

In the entertainment sector, AI is playing a pivotal role in content creation, recommendation systems, and audience engagement. Streaming services like Netflix and Spotify use AI algorithms to recommend content to users, greatly influencing viewing and listening habits. The work of Gomez-Uribe and Hunt (2016) delves into the algorithms behind these recommendation systems, revealing the complex interplay of user data and content curation [365].

AI in Content Creation and Production.

AI is not just curating content but also creating it. In the music, film, and gaming industries, AI is being used to compose music, script movies, and design game environments. This intersection of AI and creativity is pushing the boundaries of traditional content creation, as noted by Engel et al. (2017) in their study on neural networks for music composition [366].

Ethical and Social Implications.

The increasing reliance on AI in news and entertainment raises several ethical and social concerns. Issues such as data privacy, intellectual property rights, and the homogenization of content are at the forefront of discussions. Van Dijck (2013) addresses these concerns, particularly the ethical challenges posed by data-driven media platforms [367].

5.2.2 Implications for Knowledge Sharing in the Workplace

Enhancing Knowledge Dissemination with AI.

The digitization of media and the advent of AI technologies have significantly influenced knowledge sharing within professional settings. AI systems facilitate the collection, analysis, and dissemination of information, thereby enhancing organizational learning and decision-making processes. Wenger (1998) discusses the concept of communities of practice, emphasizing how digital tools can foster knowledge sharing within organizations [368].

AI-Driven Collaboration Tools.

Collaboration tools powered by AI, such as intelligent project management software and automated knowledge bases, are transforming how teams interact and share information. These tools not only streamline communication but also provide personalized insights and recommendations, thereby improving efficiency and productivity. Bughin et al. (2018) explore the impact of AI on collaboration and teamwork, highlighting the potential for AI to augment human capabilities in the workplace [369].

Challenges in AI-Enabled Knowledge Sharing.

While AI offers numerous benefits for knowledge sharing, it also presents challenges. One of the key issues is ensuring the accuracy and reliability of the information disseminated by AI systems. Furthermore, there are concerns about maintaining confidentiality and data security in AI-driven knowledge management systems. Mayer-Schönberger and Cukier (2013) delve into the challenges of big data management, including issues related to data privacy and governance [370].

The Role of AI in Organizational Learning.

AI is playing a crucial role in advancing organizational learning. By analyzing patterns in data, AI can identify gaps in knowledge and suggest areas for skill development. This ability to tailor learning and development initiatives to individual and organizational needs is revolutionizing how organizations approach talent management. Argyris and Schön (1978) provide a foundational framework on organizational learning, which is increasingly relevant in the context of AI integration [371].

Future Trends in Workplace Knowledge Sharing.

Looking ahead, the integration of AI in knowledge sharing is expected to become more sophisticated, with advancements in natural language processing and machine learning leading to more intuitive and effective knowledge management systems. The work of Davenport and Ronanki (2018) on cognitive technologies offers insights into future trends and the evolving landscape of AI in the workplace [372].

5.3 AI's Role in Creating the AI-Enabled Leisure Society

5.3.1 *Concept and Realization of Increased Leisure Time*

The Emergence of the AI-Enabled Leisure Society.

The concept of an AI-Enabled Leisure Society is rooted in the idea that advancements in artificial intelligence and automation will significantly reduce the need for human labor in various sectors. This paradigm shift suggests a future where the focus shifts from work to leisure, enabling individuals to pursue personal interests and passions. Keynes (1930) was among the first to envision this future, predicting a significant reduction in working hours due to technological progress [373].

AI's Contribution to Increased Leisure Time.

Artificial intelligence is at the forefront of this transition, automating routine tasks, optimizing workflows, and improving productivity. This automation not only streamlines work processes but also opens the possibility for shorter workweeks and longer leisure time. Brynjolfsson and McAfee (2014) discuss how AI and robotics are reshaping the economy, potentially leading to an era of unprecedented free time for individuals.

Realization in Contemporary Society.

While the concept of an AI-Enabled Leisure Society is appealing, its realization in contemporary society presents various challenges. The current economic structures, societal norms, and the distribution of the benefits of AI and automation play a significant role in determining how this free time is allocated and experienced. Rifkin (1995) explores these challenges, focusing on the distribution of wealth and time in the emerging automated economy [374].

Implications for Work-Life Balance.

The potential increase in leisure time also has profound implications for work-life balance. The redistribution of time between work and leisure could lead to a more holistic approach to life, emphasizing personal development, community engagement, and well-being. Schor (1992) investigates the relationship between work time and leisure time, emphasizing the need for a balance that promotes overall life satisfaction [375].

Technological Determinism and Societal Change.

The concept of an AI-Enabled Leisure Society also raises questions about technological determinism and its impact on societal change. While technology can drive societal transformations, the direction and nature of these changes are also shaped by cultural, economic, and political factors. Winner (1986) discusses the interplay between technology and society, cautioning against a deterministic view of technological progress [376].

5.3.2 Economic and Societal Implications

Economic Impact of an AI-Enabled Leisure Society.

The transition towards an AI-Enabled Leisure Society carries significant economic implications. One of the primary considerations is the redistribution of wealth and resources in a society where labor is not the central means of economic contribution. Autor et al. (2003) explore the impacts of technological advancements on labor markets, highlighting the changing nature of job roles and income distribution [377]. Further, Acemoglu and Restrepo (2018) delve into the economic effects of automation, examining how it influences wages and employment [378].

Societal Shifts and the Role of Policy.

Societal adaptation to an era of increased leisure time necessitates careful policy planning. Issues such as universal basic income, redefinition of work, and support for lifelong learning become crucial in this context. Bregman (2017) discusses the potential of universal basic income as a response to automation-induced job displacement [379]. Additionally, the role of education and training in preparing for a future with different labor demands is explored by Brynjolfsson et al. (2015) [380].

Changing Nature of Work and Leisure.

As AI reshapes the economic landscape, the distinction between work and leisure may become increasingly blurred. The rise of gig economies, remote work, and flexible job roles are indicative of this change. De Stefano (2016) examines the implications of the gig economy for labor laws and worker protections [381]. The impact of these changes on individual identity and societal structures is further analyzed by Huws et al. (2019) in their study of the future of work [382].

Ethical Considerations and Social Equity.

The ethical considerations surrounding an AI-Enabled Leisure Society are complex. Questions regarding social equity, access to technology, and the fair distribution of economic benefits are paramount. Susskind and Susskind (2015) address the ethical challenges of technological unemployment, emphasizing the need for equitable access to the benefits of AI [383]. Furthermore, the societal implications of AI and its impact on human dignity and social structures are discussed by Bostrom and Yudkowsky (2014) [29].

Future Outlook and Sustainable Development.

As society navigates towards an AI-Enabled Leisure Society, sustainable development and responsible use of technology become critical. The alignment of AI development with sustainable economic and social goals is essential. Schwab (2018) discusses the role of AI in achieving sustainable development goals, underscoring the importance of aligning technological advancements with societal needs [384].

5.4   The Emergence of AI in the Sharing Economy

5.4.1   *Transforming Traditional Economic Models with AI*

Redefining Economic Interactions with AI.

The sharing economy, characterized by peer-to-peer exchanges and collaborative consumption, is undergoing a transformation with the integration of AI. AI-driven platforms are redefining how goods and services are exchanged, creating new economic models that challenge traditional market structures. Sundararajan (2013) explores this shift, highlighting how digital platforms facilitated by AI are altering the dynamics of supply and demand [385].

AI in Optimizing Sharing Economy Platforms.

AI technologies play a critical role in optimizing operations within the sharing economy. Through sophisticated algorithms, AI enhances user experience by personalizing services, predicting demand, and efficiently matching supply with consumer needs. Malhotra and Van Alstyne (2014) discuss the role of digital platforms in the sharing economy, emphasizing the importance of AI in managing these platforms [386].

Economic Benefits and Scalability.

AI-driven sharing economy platforms offer significant economic benefits, including increased efficiency, reduced costs, and the creation of new market opportunities. These platforms are not limited by traditional constraints of scale, allowing for rapid growth and expansion. Zervas et al. (2017) analyze the economic impact of sharing economy platforms, particularly in the context of hospitality and transportation sectors [387].

Impact on Labor and Employment.

The rise of the AI-driven sharing economy has profound implications for labor and employment. While these platforms create new job opportunities, they also pose challenges in terms of job security, benefits, and workers' rights. Hall and Krueger (2018) examine the labor market effects of the sharing economy, focusing on the gig economy and its implications for workers [388].

Regulatory Challenges and Future Directions.

As AI reshapes the sharing economy, regulatory challenges emerge. Governments and policymakers are grappling with issues related to taxation, consumer protection, and labor rights in this new economic landscape. Cohen and Sundararajan (2015) discuss the regulatory responses to the sharing economy, suggesting paths for future policy development [389].

5.4.2 Case Studies: AI-Driven Sharing Platforms

Airbnb: Revolutionizing Hospitality with AI.

Airbnb, a leading player in the sharing economy, utilizes AI to enhance user experience and operational efficiency. AI algorithms on Airbnb's platform assist in price optimization, personalized recommendations, and fraud detection. Guttentag (2015) examines the impact of Airbnb on the traditional hotel industry, noting the role of AI in Airbnb's business model [390].

Uber: Transforming Urban Transportation.

Uber's success in revolutionizing urban transportation is partly attributed to its use of AI. The platform's AI algorithms optimize routes, match riders with drivers, and predict demand patterns. Cramer and Krueger (2016) analyze the economic effects of Uber and similar platforms on the transportation industry, highlighting the role of AI in their operational models [391].

TaskRabbit: AI in the Gig Economy.

TaskRabbit, a platform connecting freelancers with local demand for everyday tasks, leverages AI for task matching and pricing strategies. The use of AI allows for efficient market functioning and enhances user satisfaction. Todolí-Signes (2017) discusses the implications of such gig economy platforms on labor laws and the workforce [392].

AI-Driven Platforms in Healthcare: Zocdoc and Babylon Health.

In the healthcare sector, platforms like Zocdoc and Babylon Health utilize AI for appointment scheduling, medical consultations, and personalized health recommendations. These platforms demonstrate AI's potential in improving healthcare accessibility and efficiency. Blease et al. (2018) explore the role of digital platforms in healthcare, focusing on patient engagement and service delivery [393].

Implications for Industry and Society.

These case studies highlight AI's transformative role in various industries within the sharing economy. The integration of AI not only improves business operations but also has broader implications for industry structures and societal dynamics. Sundararajan (2016) provides a comprehensive analysis of the sharing economy's impact on society, including the role of AI-driven platforms [394].

5.5 AI, Robotics, and Automation in the Workplace

5.5.1 Sector-Specific Transformations

Manufacturing: Automation and Efficiency.

In the manufacturing sector, robotics and AI have revolutionized production processes. Automation has led to increased efficiency, precision, and safety, while AI optimizes supply chains and predictive maintenance. Bessen (2016) explores the impact of automation on manufacturing jobs, emphasizing both the displacement and creation of new types of employment [395].

Healthcare: AI-Enhanced Diagnostics and Treatment.

The healthcare sector is experiencing significant transformation with the integration of AI in diagnostics, treatment planning, and patient care management. AI algorithms aid in early detection of diseases and help in formulating personalized treatment plans. Obermeyer and Emanuel (2016) discuss the role of AI in healthcare, particularly in improving diagnostic accuracy and treatment outcomes [396].

Retail: Personalization and Inventory Management.

In retail, AI and automation are used for personalizing customer experiences, managing inventory, and optimizing logistics. AI-driven analytics provide insights into consumer behavior, enhancing the efficiency of supply chain operations. Grewal et al. (2017) analyze the influence of AI on retail, focusing on customer engagement and operational improvements [397].

Finance: AI in Risk Assessment and Decision Making.

The financial sector is leveraging AI for risk assessment, fraud detection, and automated trading. AI algorithms analyze market trends and customer data to make informed financial decisions. Arner et al. (2016) examine the integration of AI in financial services, highlighting its impact on financial markets and institutions [398].

Agriculture: Robotics in Farming and Crop Management.

AI and robotics are transforming agriculture through precision farming, automated equipment, and data-driven crop management. These technologies enable more efficient use of resources, higher yields, and sustainable farming practices. Liakos et al. (2018) investigate the application of AI in agriculture, noting its potential to revolutionize farming practices [399].

5.5.2 Human-AI Collaboration and New Job Roles

The Synergy of Human and AI Collaboration.

The integration of AI in the workplace is not just about automation; it's also about augmenting human capabilities. This collaboration between humans and AI is leading to enhanced productivity and creativity. Daugherty and Wilson (2018) discuss the complementary nature of human-AI collaboration, emphasizing how AI can augment human skills rather than replace them [241].

Emergence of New Job Roles.

As AI takes on more routine tasks, new job roles are emerging that focus on AI management, development, and ethical oversight. These roles require a blend of technical skills and understanding of AI's broader implications. Kaplan and Haenlein (2019) explore the evolving job landscape, noting the creation of roles such as AI trainers, explainers, and sustainers [354].

Reskilling and Upskilling the Workforce.

The shift towards an AI-integrated workplace necessitates reskilling and upskilling of the current workforce. Organizations are increasingly investing in training programs to prepare employees for the changing job requirements. Faggella (2018) highlights the importance of education and training in enabling workers to adapt to AI-driven changes [400].

AI in Creative and Cognitive Tasks.

Contrary to the belief that AI is limited to routine tasks, it is also making inroads into creative and cognitive domains. AI's role in assisting with creative processes and decision-making is expanding, leading to new collaborative opportunities in fields like design, marketing, and strategy. Broussard (2018) examines AI's involvement in creative tasks, emphasizing its potential to enhance human creativity [401].

Ethical and Societal Considerations in Human-AI Collaboration.

As AI becomes a more integral part of the workforce, ethical and societal considerations gain prominence. Issues around transparency, accountability, and bias in AI decision-making are increasingly relevant. Crawford and Calo (2016) address the ethical challenges in human-AI collaboration, advocating for responsible and equitable AI development [402].

5.6 Technological Unemployment and Job Displacement

5.6.1 Assessing the Risks and Opportunities

Understanding Technological Unemployment.

Technological unemployment, a phenomenon where jobs are lost due to technological advancements, is a growing concern in the AI era. The risk of job displacement varies across sectors and job types, with routine and manual tasks being more susceptible to automation. Frey and Osborne (2017) provide a comprehensive analysis of the susceptibility of jobs to computerization, highlighting significant disparities across different sectors [48].

Economic Impact of Job Displacement.

The economic impact of technological unemployment extends beyond individual job losses. It encompasses changes in wage structures, shifts in labor market dynamics, and potential increases in income inequality. Acemoglu and Restrepo (2020) explore the broader economic implications of automation, including its effects on wages and employment [262].

Opportunities in a Technologically Advanced Economy.

Despite the risks, technological advancements also create new opportunities. New job roles are emerging, particularly in AI development, data analysis, and technology ethics. Bessen (2019) discusses how technological change often leads to the creation of new types of jobs, emphasizing the importance of adaptability and skill development [403].

The Role of Education and Policy in Mitigating Risks.

To mitigate the risks of technological unemployment, focus on education, skill development, and policy interventions is crucial. Upskilling and reskilling initiatives are key to preparing the workforce for a changing job landscape. Autor (2015) addresses the role of education and training in adapting to technological changes, suggesting policies to support workforce transitions [404].

Shaping the Future Workforce.

The future workforce must be equipped to navigate the challenges and opportunities presented by AI and automation. Collaborative efforts among governments, educational institutions, and businesses are essential in shaping a workforce that is resilient to technological disruptions. Schwab (2018) discusses strategies for preparing for the Fourth Industrial Revolution, focusing on collaborative and proactive approaches [384].

5.6.2   Strategies for Workforce Transition

Identifying Future Skill Requirements.

A critical strategy for workforce transition is identifying the skills that will be in high demand in an AI-driven economy. This involves analyzing emerging trends in AI and automation to predict future job roles and required competencies. Bughin et al. (2018) emphasize the importance of forecasting skill requirements to prepare the workforce for future challenges [405].

Upskilling and Reskilling Initiatives.

Upskilling and reskilling are pivotal in equipping the current workforce with the skills needed in a technologically advanced job market. These initiatives should focus on digital literacy, problem-solving, and adaptability. McKinsey Global Institute (2017) provides insights into the scale of retraining needed and the types of skills that will be essential in the future [369].

Public-Private Partnerships in Education and Training.

Collaboration between public and private sectors is key to effective workforce transition strategies. Partnerships can facilitate the development of training programs that are aligned with industry needs and technological advancements. Brynjolfsson et al. (2017) discuss the role of public-private partnerships in driving effective workforce development [406].

Policy Interventions and Support Systems.

Government policy plays a crucial role in supporting workforce transitions. This includes unemployment benefits, job search assistance, and incentives for companies to invest in employee training. Autor and Dorn (2013) examine the impact of policy interventions on labor market adjustments to technological changes [407].

Cultivating a Culture of Lifelong Learning.

Fostering a culture of lifelong learning is essential for workforce adaptability in an ever-evolving technological landscape. Continuous learning ensures that the workforce remains agile and can cope with rapid changes in job requirements. World Economic Forum (2018) highlights the importance of lifelong learning in ensuring workforce resilience and adaptability [408].

5.7 Education and Skill Development in an AI Era

5.7.1 *Adapting Educational Systems for AI Integration*

Rethinking Education in the AI Era.

The integration of AI into the economy necessitates a fundamental rethinking of educational systems. Traditional education models, which have long emphasized rote learning and standardized testing, are becoming increasingly inadequate for preparing students for a future dominated by AI and automation. The shift calls for an educational paradigm that prioritizes critical thinking, creativity, and digital literacy. Wagner and Dintersmith (2015) argue for a transformative change in education, focusing on fostering skills like collaboration, communication, and problem-solving, which are crucial in an AI-driven world [409]. Moreover, the integration of AI tools in the learning process itself can provide personalized learning experiences, adapt to individual student needs, and enhance the efficiency and effectiveness of education. Luckin et al. (2016) explore the potential of AI in education, highlighting how AI can transform teaching methodologies and learning outcomes [410].

The need for continuous learning and adaptability is more pronounced than ever, as the half-life of skills is rapidly shrinking in the face of technological advancements. Educational institutions must therefore pivot towards curricula that are more aligned with the evolving job market, emphasizing digital skills and AI literacy. A report by the World Economic Forum (2018) underscores the urgency of revising educational curricula to include skills that are relevant to the Fourth Industrial Revolution [411]. Furthermore, partnerships between educational institutions and industry are critical in ensuring that the skills taught are directly applicable to the workforce needs. Bessen (2017) discusses the importance of such collaborations, illustrating how they can bridge the gap between education and employment [412].

Beyond formal education, there is a growing need for informal and lifelong learning pathways. Online platforms and MOOCs (Massive Open Online Courses) are playing an increasingly vital role in providing accessible education to a broader audience. These platforms not only democratize education but also provide flexibility for individuals to acquire new skills as per the market demands. Agarwal (2014) emphasizes the role of online education in democratizing access to quality education, making it a key player in the global education landscape [413]. Additionally, government policies play a crucial role in shaping education systems for the AI era. Policymakers must recognize the changing landscape and support initiatives that promote AI literacy, from primary education to adult learning programs. Goldin and Katz (2018) delve into the policy implications for education in an age of technological change, advocating for government intervention to ensure equitable access to education [414].

5.7.2 Lifelong Learning and Reskilling Initiatives

Embracing Lifelong Learning in an AI-Driven World.

The rapid advancement of AI and automation technologies necessitates a shift towards lifelong learning, where continuous skill development becomes an integral part of one's career. In an era where the half-life of professional skills is shrinking, the ability to adapt and learn new skills is paramount. This paradigm shift requires a cultural change where learning is seen as a continuous journey rather than a one-time endeavor. The concept of lifelong learning is gaining traction, as it enables workers to stay relevant and competitive in an ever-evolving job market. Fosway Group (2019) emphasizes the importance of lifelong learning in the context of digital transformation, highlighting its role in maintaining workforce agility and adaptability [415]. As AI continues to reshape industries, traditional job roles are evolving, and entirely new professions are emerging, necessitating a reassessment of skill sets and learning methodologies. The European Commission (2020) addresses this phenomenon, advocating for robust lifelong learning policies to support workforce transitions in the digital age [416].

The development of reskilling initiatives is crucial for workforce adaptation to the changes brought about by AI and automation. These initiatives must not only focus on technical skills pertinent to AI but also on soft skills like problem-solving, critical thinking, and emotional intelligence, which are equally vital in the AI era. Reskilling programs need to be accessible and tailored to the needs of diverse learners, including those already in the workforce and those entering the job market. The World Economic Forum (2020) provides insights into successful reskilling models, outlining strategies to upskill the global workforce effectively [417]. Collaboration between governments, educational institutions, and private organizations is key to developing comprehensive reskilling programs. Such collaborations can ensure that the programs are aligned with current and future job market demands, thereby enhancing their effectiveness. Brynjolfsson et al. (2020) explore the role of public-private partnerships in facilitating large-scale skill development initiatives, emphasizing the need for coordinated efforts to address the skills gap [418].

In addition to structured reskilling programs, the rise of digital platforms and online learning resources has made education more accessible than ever. MOOCs, e-learning platforms, and virtual training programs offer flexible and personalized learning experiences, enabling individuals to learn at their own pace and according to their unique career paths. These platforms are not only democratizing education but also providing opportunities for continuous learning outside traditional classroom settings. Reich and Ruipérez-Valiente (2019) examine the impact of online learning platforms on education, noting their potential to support lifelong learning and professional development [419]. Governments play a pivotal role in fostering a culture of lifelong learning, through policies that support educational access, incentivize reskilling, and recognize the importance of non-traditional learning pathways. The OECD (2019) discusses policy approaches to lifelong learning, underscoring the need for governmental support in creating a resilient and adaptable workforce [420].

In conclusion, as AI continues to transform the workplace, lifelong learning and reskilling initiatives are becoming increasingly crucial. These initiatives must be comprehensive, inclusive, and aligned with future workforce needs, ensuring that individuals are prepared to thrive in an AI-driven world. The collaborative efforts of various stakeholders, including governments, educational institutions, and private organizations, are essential in realizing the full potential of these initiatives.

5.7.3   AI Fairness and Bias

The Challenge of AI Fairness and Bias in the Workplace.

The integration of AI in the workplace has raised significant ethical concerns, with fairness and bias being at the forefront. AI systems, while offering numerous benefits in terms of efficiency and decision-making, can also perpetuate and amplify existing biases if not carefully designed and monitored. The potential for AI to inherit prejudices from biased training data or to develop biased algorithms due to flawed design is a major concern. Barocas and Selbst (2016) delve into the issue of algorithmic fairness and discrimination, discussing how AI systems can inadvertently encode and perpetuate societal biases [421]. This issue is particularly pertinent in areas such as recruitment, performance evaluations, and promotion, where AI-driven decisions can have significant impacts on individuals' careers and livelihoods. Eubanks (2018) provides insights into the impact of automated decision-making in public services, highlighting cases where biased algorithms led to unfair treatment [260].

The responsibility to ensure AI fairness in the workplace lies with both the developers of AI systems and the organizations that employ them. There is a growing recognition of the need for ethical AI frameworks and guidelines that can help in mitigating bias. Initiatives such as the AI Now Institute's work focus on developing more equitable and accountable AI systems. Moreover, transparency in AI algorithms and decision-making processes is crucial for ensuring fairness. This involves making AI systems' workings understandable to non-experts, allowing for scrutiny and accountability. Burrell (2016) examines the challenges of interpreting AI systems, advocating for greater transparency and interpretability in AI design [422].

Diversity in AI development teams is another key factor in combating bias in AI systems. A diverse group of developers is more likely to recognize and address potential biases in AI systems. Buolamwini and Gebru (2018) emphasize the importance of inclusive and diverse teams in AI development, showcasing how lack of diversity can lead to biased facial recognition technologies [41]. Additionally, continuous monitoring and auditing of AI systems in the workplace are essential for identifying and rectifying biases. Regular audits can help ensure that AI systems operate fairly and do not perpetuate discrimination. Raji et al. (2020) discuss the need for ongoing evaluation of AI systems, proposing frameworks for auditing AI for bias and fairness [197].

In conclusion, addressing fairness and bias in AI is a multifaceted challenge that requires concerted efforts from various stakeholders. It involves careful design and development of AI systems, transparency and interpretability in AI operations, diversity in development teams, and ongoing monitoring and auditing of AI systems. By tackling these issues, organizations can harness the benefits of AI in the workplace while upholding ethical standards and ensuring fair treatment of all employees.

5.7.4 Privacy and Data Security Concerns

Navigating Privacy in the AI-Driven Workplace.

The proliferation of AI in the workplace brings with it significant concerns regarding privacy and data security. As AI systems increasingly handle large volumes of personal and sensitive data, the risk of breaches and misuse of this data escalates. Protecting employee privacy becomes a paramount concern, especially when AI is used in monitoring, performance evaluation, and personal data analysis. Cavoukian and Jonas (2012) discuss the concept of privacy by design, emphasizing the need to incorporate privacy protections into the development of AI technologies [423]. The use of AI in the workplace to track employee activities, analyze behavior, and predict performance raises ethical questions about the extent to which monitoring is justifiable. Wachter (2018) explores the balance between data-driven innovation and the protection of individual privacy, highlighting legal and ethical frameworks that govern data use in the workplace [75].

The increasing sophistication of AI systems also amplifies concerns about data security. The potential for AI to be used in cyberattacks or for AI systems to be compromised poses a threat not only to individual privacy but also to corporate and national security. As AI becomes more integrated into business operations, the importance of robust cybersecurity measures grows. Taddeo and Floridi (2018) analyze the ethical implications of AI in cybersecurity, discussing the dual role of AI as both a tool for and a target of cyber threats [424]. Furthermore, the collection and storage of large datasets by AI systems necessitate stringent data governance policies. Policies need to address aspects such as data access, consent, and the right to be forgotten, ensuring that data is used responsibly and ethically. Mayer-Schönberger (2018) emphasizes the significance of data governance in the age of AI, advocating for regulations that protect individual rights while enabling innovation [425].

Employee trust is a critical factor in the successful implementation of AI in the workplace. Transparent policies regarding data collection, usage, and security are essential to build and maintain this trust. Employees need to be assured that their data is handled securely and used in a manner that respects their privacy. Martin (2019) highlights the importance of trust in the digital age, suggesting that organizations must be transparent and accountable in their use of AI and data [426]. In addition to internal policies, external regulations play a crucial role in ensuring data privacy and security. The General Data Protection Regulation (GDPR) in the European Union is an example of a regulatory framework that sets standards for data protection, impacting how AI is used in workplaces globally. Voigt and Von dem Bussche (2017) provide an overview of GDPR and its implications for AI and data privacy [427].

In conclusion, privacy and data security are key ethical considerations in the deployment of AI in the workplace. Addressing these concerns involves a combination of privacy-by-design approaches in AI development, robust data governance policies, transparency with employees, and adherence to regulatory standards. Ensuring the ethical use of AI and the protection of personal data is essential for maintaining trust and integrity in AI-driven workplaces.

5.8 The Digital Divide and Inclusivity in the AI Workplace

5.8.1 Addressing Inequalities in Access and Skills

The Digital Divide in the AI Era.

The advent of AI in the workplace has exacerbated the digital divide, a term that refers to the disparity between those who have access to modern information and communication technology and those who do not. This divide is not just about access to technology, but also encompasses the skills needed to effectively use AI and digital tools. The digital divide has significant implications for inclusivity in the AI workplace, as it can lead to unequal opportunities and reinforce existing socio-economic disparities. Van Dijk (2017) discusses the various dimensions of the digital divide, emphasizing the need for comprehensive strategies to bridge these gaps [428]. In the context of the workplace, the divide manifests in disparities in AI literacy and access to AI-enhanced tools, which can affect career advancement and job security. Robinson et al. (2015) explore the impact of the digital divide on employment, highlighting how lack of access to technology can limit job opportunities and career growth [429].

Bridging the AI Skill Gap.

To address the digital divide in the workplace, it is crucial to focus on bridging the AI skill gap. This involves not only providing access to technology but also ensuring that individuals have the necessary skills to use AI tools effectively. Education and training programs need to be inclusive, catering to diverse demographics and backgrounds. Such programs should aim to equip individuals with a blend of technical, analytical, and soft skills that are essential in an AI-driven job market. Rainie and Anderson (2017) highlight the importance of lifelong learning and adaptability in an era where AI and automation are rapidly changing job requirements [430]. Additionally, employers play a critical role in providing training and development opportunities to their workforce, ensuring that employees are not left behind in the technological shift. Atkinson and Wu (2017) emphasize the responsibility of businesses in facilitating skill development, advocating for corporate investment in employee training [431].

Inclusivity and Equity in AI Deployment.

Inclusivity in the AI workplace goes beyond access to technology and encompasses the creation of an equitable work environment. This includes addressing biases in AI algorithms that may lead to discriminatory practices in hiring, promotions, and performance evaluations. Ensuring that AI tools are designed and implemented in a way that is fair and unbiased is crucial for building an inclusive workplace. Eubanks (2018) examines the ethical considerations in AI deployment, focusing on how AI can be used responsibly to avoid reinforcing existing inequalities [260]. Furthermore, fostering a workplace culture that values diversity and inclusivity is essential in realizing the full potential of AI. Such a culture encourages diverse perspectives in AI development and decision-making, leading to more robust and equitable AI solutions. D'Ignazio and Klein (2020) discuss the significance of diversity in data science and AI, arguing for the inclusion of marginalized voices in technology design and implementation [432].

In conclusion, addressing the digital divide and promoting inclusivity in the AI workplace are multifaceted challenges that require concerted efforts from various stakeholders. These efforts include providing equitable access to technology and skills, fostering lifelong learning, ensuring equitable AI deployment, and cultivating an inclusive workplace culture. By tackling these challenges, organizations can harness the benefits of AI while ensuring fairness and equity for all employees.

5.8.2 Policies for Inclusive AI Development

Crafting Policies for Equitable AI Integration.

The development and integration of AI in the workplace necessitate thoughtful policy-making to ensure inclusivity and equity. Policymakers face the challenge of creating frameworks that not only foster AI innovation but also protect the interests of all stakeholders, particularly those who might be disadvantaged by the digital divide. This task involves regulating AI deployment in a way that promotes fair access and prevents discrimination. The European Commission's ethics guidelines for trustworthy AI offer an example of efforts to ensure AI systems are lawful, ethical, and robust [433]. Policies must also address the broader implications of AI on the labor market, ensuring that transitions due to automation are managed in a socially responsible manner. The OECD's principles on AI emphasize the importance of inclusive growth, sustainable development, and well-being [434].

Equally important is the role of educational policies in preparing the current and future workforce for an AI-driven economy. This involves not only incorporating AI and digital literacy into curricula but also promoting lifelong learning and continuous skill development. The focus should be on both technical skills relevant to AI and soft skills like critical thinking and adaptability. The UNESCO's Beijing Consensus on Artificial Intelligence and Education outlines strategies for leveraging AI to enhance education while addressing ethical and equity concerns [435].

Corporate policies also play a critical role in ensuring inclusive AI development within organizations. Businesses need to commit to diversity and inclusivity in their AI strategies, including diverse representation in AI development teams and decision-making processes. This approach helps in recognizing and mitigating biases in AI systems. The Partnership on AI's Tenets provide guidance on responsible AI development and use, focusing on safety, fairness, and accountability [436].

In tandem, there's a need for policies that support workers displaced by AI and automation. Such policies could include job transition programs, unemployment benefits tailored to the gig economy, and incentives for businesses to retrain and redeploy workers affected by automation. Brynjolfsson and McAfee (2014) discuss the broader economic implications of AI, suggesting policy interventions to ensure equitable benefits [291].

Ultimately, the goal of policy-making in the realm of AI should be to harness its potential while safeguarding against its risks, ensuring that the benefits of AI are distributed equitably across society. The World Bank's report on the changing nature of work underlines the need for an inclusive approach to technological advancement, highlighting policies that can help in creating a more equitable future of work [437]. By balancing innovation with ethical considerations and equity, policies can pave the way for AI to be a positive force in the workplace and beyond.

5.9 Regulating AI in the Work Environment

5.9.1 National and International Policy Frameworks

The Role of National Regulations in AI Governance.

The governance of AI in the workplace is increasingly becoming a priority at the national level, with various countries implementing policies and regulations to guide the ethical and effective use of AI technologies. These national frameworks often focus on promoting innovation while ensuring that AI development aligns with societal values and labor laws. In the United States, initiatives like the AI in Government Act reflect an increasing focus on establishing guidelines for AI use in public sectors and workplaces [438]. Similarly, the European Union's AI strategy encompasses comprehensive regulations that address AI's impact on labor, privacy, and ethical standards, setting a precedent for responsible AI development and deployment [76].

The intricacies of regulating AI in the workplace include considerations around data privacy, employee surveillance, decision-making transparency, and ensuring that AI-driven decisions do not lead to discrimination or unfair treatment. National frameworks must balance the need for innovation with worker protections and ethical considerations. The UK's House of Lords report on AI emphasizes the importance of ethical AI development and the need for AI to work for the public good [439]. In Asian countries, such as Japan and South Korea, government policies are increasingly focusing on integrating AI in industries while addressing the social and ethical implications of this integration [440].

International Cooperation and Policy Harmonization.

As AI technologies transcend borders, international cooperation becomes essential in developing consistent and effective policy frameworks. International bodies like the United Nations and the OECD play a crucial role in fostering dialogue and collaboration among countries. The OECD's AI Principles, adopted by numerous countries, provide a global reference point for trustworthy AI, emphasizing principles like transparency, fairness, and accountability [74]. These international guidelines help in harmonizing AI policies across borders, ensuring a cohesive approach to AI governance.

In addition to intergovernmental efforts, collaborations among international organizations, academia, and the private sector are crucial in shaping global AI policies. These collaborations can lead to the development of standards and best practices that guide AI deployment in diverse work environments. The IEEE's Global Initiative on Ethics of Autonomous and Intelligent Systems is an example of a multi-stakeholder effort to develop ethically aligned design principles for AI [441].

Challenges and Future Directions in AI Regulation.

Despite these efforts, regulating AI in the workplace presents numerous challenges. One major challenge is keeping pace with the rapid development of AI technologies. Regulatory frameworks need to be adaptable and flexible to remain relevant as AI evolves. Another challenge is ensuring that regulations do not stifle innovation while protecting workers' rights and ethical standards. Future directions in AI regulation will likely involve continuous dialogue among policymakers, technologists, and other stakeholders, as well as periodic reviews and updates to existing policies.

Forging Ahead with Responsible AI Governance.

Ultimately, the goal of national and international AI policy frameworks is to foster a landscape where AI can be leveraged for economic growth and societal benefit, while safeguarding against potential risks and ethical dilemmas. This balance is crucial for the sustainable and responsible development of AI technologies in the workplace. As nations and international bodies forge ahead with AI governance, the focus will be on creating agile, inclusive, and ethically grounded frameworks that can adapt to the evolving landscape of AI and its implications for the global workforce.

5.9.2 Best Practices for AI Governance in Organizations

Establishing Ethical AI Frameworks in Organizations.

As organizations increasingly adopt AI technologies, establishing robust governance frameworks becomes crucial to ensure ethical, transparent, and accountable AI use. The development of these frameworks involves setting clear guidelines on AI ethics, data handling, privacy, and decision-making processes. Developing a corporate AI ethics policy is a fundamental step, outlining the organization's commitment to ethical AI use. Such policies should reflect principles like fairness, accountability, and transparency, ensuring AI systems align with the organization's values and legal standards. Jobin et al. (2019) analyze various AI ethics guidelines, highlighting common themes and practices for ethical AI deployment [47].

A key aspect of AI governance is ensuring transparency in AI systems. This involves not only making the workings of AI algorithms understandable but also ensuring that decisions made by AI systems can be explained and justified. Transparency is essential for building trust among employees and stakeholders and for complying with regulatory requirements. Floridi et al. (2018) discuss the importance of transparent AI, emphasizing its role in ensuring responsible AI usage [11].

Another best practice is the implementation of continuous monitoring and auditing of AI systems. Regular audits help identify and mitigate biases, inaccuracies, or unethical practices in AI systems. Such audits should be conducted by interdisciplinary teams, including ethicists, data scientists, and legal experts, to ensure a comprehensive evaluation. Rahwan (2018) explores the concept of society-in-the-loop, advocating for ongoing societal and ethical audits of AI systems [442].

Engaging a diverse set of stakeholders in the development and deployment of AI is also crucial for effective governance. Diversity in AI development teams can help prevent biases in AI systems and ensure that a wide range of perspectives is considered. Diversity not only refers to demographic diversity but also to diversity in expertise and viewpoints. West et al. (2019) highlight the significance of diversity and inclusion in AI development, pointing out how it leads to more robust and equitable AI solutions [320].

In addition, organizations should focus on building AI literacy among their workforce. Educating employees about AI, its capabilities, and its limitations empowers them to engage with AI systems more effectively and raises awareness about ethical considerations. This approach also helps in demystifying AI, reducing fears and misconceptions about AI taking over jobs. Tegmark (2017) discusses the need for widespread AI education, emphasizing its role in preparing society for an AI-driven future [443].

Synthesizing Ethical AI Practices for Organizational Success.

In synthesizing these practices, it's evident that successful AI governance in organizations requires a multifaceted approach. It involves establishing ethical frameworks, ensuring transparency and explainability, conducting regular audits, promoting diversity and inclusivity, and fostering AI literacy among employees. By adhering to these best practices, organizations can leverage AI's potential while maintaining ethical integrity and trust, ultimately contributing to a sustainable and responsible AI-driven work environment.

5.10   AI and the Global Economy

5.10.1   *AI's Impact on Global Economic Structures*

Transforming Economic Landscapes with AI.

Artificial Intelligence is playing a pivotal role in transforming global economic structures, leading to significant shifts in how economies function and compete. AI's ability to process vast amounts of data, automate complex tasks, and optimize decision-making processes is changing the nature of work, production, and market dynamics. This transformation is not limited to advanced economies; it's impacting nations at all stages of development. Baldwin (2019) discusses how AI and digital technology are altering global economic patterns, reducing the importance of geographical proximity in economic activities [444]. In many industries, AI-driven automation is leading to increased productivity but also raising concerns about job displacement and economic inequality. Brynjolfsson et al. (2018) examine the dual effects of AI on productivity and employment, highlighting both the challenges and opportunities it presents for economic growth [445].

AI's Role in Global Trade and Competitiveness.

AI is also reshaping global trade, influencing how countries engage in and benefit from international markets. Advanced AI capabilities can provide a significant competitive advantage in global trade, influencing everything from manufacturing and services to agriculture and healthcare. Countries investing heavily in AI research and development are poised to gain a substantial edge in global markets. Lee (2018) analyzes the race for AI dominance, noting the strategic importance of AI in national competitiveness [22]. Furthermore, AI is enabling more efficient supply chain management, transforming global logistics, and trade patterns. Muro et al. (2019) explore how AI-driven automation is affecting global supply chains, impacting labor markets and trade policies [356].

Economic Policy and AI Governance on a Global Scale.

The widespread impact of AI on the global economy necessitates thoughtful economic policies and international governance frameworks. Policymakers face the challenge of harnessing AI's potential for economic growth while mitigating its disruptive effects on labor markets and societal structures. International cooperation is essential to address these challenges, as AI's impact transcends national borders. Schwab (2017) emphasizes the need for global collaboration in shaping the Fourth Industrial Revolution, advocating for policies that promote inclusive and sustainable growth [309]. Addressing AI's impact on the global economy also involves dealing with issues of data governance, intellectual property, and the ethical use of AI. The World Economic Forum (2018) discusses these policy challenges, suggesting a framework for international collaboration in AI governance [411].

Steering Toward an Equitable AI-Driven Global Economy.

As we advance further into an AI-driven era, the global economy is at a crossroads. The path forward involves navigating the challenges posed by AI while leveraging its potential to drive innovation, efficiency, and growth. Ensuring that the benefits of AI are distributed equitably across societies and nations is paramount. The United Nations Development Programme (2019) addresses the role of AI in achieving sustainable development goals, highlighting the need for policies that ensure AI contributes to inclusive and equitable development [446]. It is imperative that global economic policies around AI are developed with a focus on equity, sustainability, and shared prosperity, ensuring that AI becomes a positive force in the global economic landscape.

In synthesizing these perspectives, it is clear that AI's impact on the global economy is profound and multifaceted. It presents opportunities for economic growth and efficiency but also poses challenges related to employment, equity, and international competitiveness. The need for proactive and collaborative policy-making at both national and international levels is crucial to ensure that AI's economic impact is positive and inclusive.

5.10.2 International Collaboration and Competition

The Dual Nature of AI in the Global Arena.

In the global economy, AI serves as both a catalyst for collaboration and a focal point for competition among nations. On one hand, international collaboration in AI research and development can lead to shared benefits, such as advancements in healthcare, environmental protection, and global security. On the other hand, AI is also a key area of strategic competition, with nations vying for technological supremacy, which has significant implications for economic power and national security. Lee (2018) delves into the competitive aspect of AI, particularly between China and the United States, highlighting how AI has become central to national strategies [22]. The role of AI in shaping international relations and global power structures is becoming increasingly pronounced, necessitating a nuanced understanding of how cooperation and competition coexist in this domain.

Frameworks for International AI Collaboration.

International collaboration in AI is facilitated through various frameworks and initiatives, involving governments, academia, and the private sector. Such collaboration aims to address global challenges that transcend national borders, including climate change, pandemic response, and sustainable development. The United Nations' AI for Good initiative exemplifies efforts to harness AI for global societal benefit, fostering collaborative projects and knowledge sharing [446]. Similarly, the Global Partnership on AI (GPAI) brings together experts from diverse fields to guide the responsible development and use of AI, emphasizing a human-centric approach [447].

Competitive Dynamics in AI Development.

While collaboration is key, AI also drives competitive dynamics among nations. Investment in AI research, development of AI industries, and the acquisition of AI talent have become vital components of national economic strategies. Countries are increasingly recognizing the importance of AI in gaining a competitive edge, leading to significant investments in AI research and development. The European Commission's investment in AI, as part of its Digital Single Market strategy, illustrates this competitive drive, aiming to position Europe at the forefront of AI innovation [76]. The competition in AI also extends to the corporate sector, where companies vie for market leadership in AI technologies, influencing global economic trends and innovation trajectories.

Balancing Competition with Cooperation.

Navigating the balance between competition and cooperation in AI is a complex but essential task. While competition can drive innovation and progress, unbridled rivalry may lead to ethical compromises, data privacy breaches, and potential misuse of AI. Establishing international norms and ethical standards for AI is crucial to ensure that competitive pursuits do not undermine global cooperation and stability. The IEEE's work on ethical AI standards is an example of efforts to establish global norms, promoting responsible AI development that benefits humanity [441].

Forging a Collaborative Future in AI.

Looking ahead, the future of AI in the global economy will likely be shaped by a combination of competitive innovation and collaborative initiatives. Nations and organizations that can effectively navigate this dual dynamic will be well-positioned to lead in the AI era. International cooperation, combined with healthy competition, can foster a diverse and robust AI ecosystem, driving technological advancements while ensuring ethical and equitable outcomes. The OECD's AI policy observatory serves as a platform for sharing best practices and policy experiences, supporting both competitive advancement and collaborative efforts in AI [74].

To summarize, the intersection of international collaboration and competition in AI presents both challenges and opportunities. It requires a careful balance to harness AI's potential for global good while acknowledging and managing the competitive nature of technological advancement. By fostering collaborative environments and establishing shared norms, the global community can ensure that AI development leads to positive outcomes for all nations.

5.11 Innovations in AI and Future Work Trends

5.11.1 *Emerging Technologies Shaping the Future of Work*

AI as a Driver of Workplace Transformation.

Artificial Intelligence is rapidly evolving, continuously introducing new technologies that are reshaping the nature of work. These innovations are not only automating routine tasks but are also enhancing human capabilities and creating new forms of collaboration between humans and machines. From intelligent automation to AI-powered analytics and decision-making tools, the scope of AI's impact is vast. Agrawal et al. (2018) discuss how AI is transforming industries by creating new opportunities for innovation and efficiency [246]. The proliferation of AI in various sectors is leading to shifts in job roles, requiring a workforce that is adaptable and skilled in new technologies.

Revolutionizing Work with Advanced AI Applications.

Advanced AI applications, such as machine learning, natural language processing, and robotics, are revolutionizing traditional work processes. In sectors like healthcare, AI is enabling precision medicine and enhancing patient care through advanced diagnostics and personalized treatment plans. In finance, AI-driven algorithms are used for risk assessment, fraud detection, and automated trading, significantly altering the financial landscape. Bostrom and Yudkowsky (2014) explore the transformative potential of AI, highlighting its capacity to redefine existing job roles and create new ones [29].

The Emergence of New Job Roles and Skills.

As AI continues to evolve, it is creating new job roles and skill demands. The rise of AI ethics officers, machine learning engineers, and data scientists reflects the growing need for professionals who can develop, manage, and oversee AI systems. These roles require a combination of technical expertise, ethical understanding, and strategic thinking. Susskind and Susskind (2015) examine the future of professions in light of AI advancements, emphasizing the changing nature of work and the skills required in the AI era [383].

Challenges and Opportunities in AI-Driven Work Environments.

While AI presents significant opportunities for enhancing work processes and creating new job roles, it also poses challenges. One of the key challenges is ensuring a smooth transition for workers whose jobs are affected by AI automation. This requires effective reskilling and upskilling programs, as well as policies that support workforce transitions. Another challenge is maintaining ethical standards and ensuring that AI systems are designed and used responsibly. Daugherty and Wilson (2018) address these challenges, outlining strategies for businesses to harness the power of AI while mitigating its risks [241].

Envisioning the Future of Work with AI.

Looking forward, the future of work in the AI era is likely to be characterized by a blend of human intelligence and machine capabilities. This integration promises to enhance creativity, problem-solving, and decision-making, leading to more innovative and efficient work environments. However, realizing this potential requires a concerted effort from businesses, governments, and educational institutions to foster a workforce that is equipped to thrive in an AI-driven world. The World Economic Forum (2020) provides insights into the future of jobs, highlighting the skills and strategies needed to navigate the evolving work landscape [408].

In sum, AI's role in shaping the future of work is multifaceted, encompassing the automation of tasks, the creation of new job roles, and the transformation of existing ones. As AI continues to advance, it offers both challenges and opportunities for the workforce. Navigating this transition effectively will require adaptive skills, continuous learning, and proactive policy-making, ensuring that the benefits of AI innovations are maximized while addressing potential risks and inequalities.

5.12 Predictions and Scenarios for the Next Fifty Years

5.12.1 Long-Term Forecasts in the AI-Driven Future

Envisioning the AI Landscape in Fifty Years.

Predicting the trajectory of AI over the next fifty years involves exploring a wide range of possibilities, from transformative advancements to potential challenges and societal shifts. AI's rapid evolution suggests a future where its integration into every aspect of human life and industry is profound. Kurzweil (2016) projects an accelerated advancement in AI, envisioning a future where AI surpasses human intelligence, leading to a paradigm shift in how we understand intelligence and its applications [448]. This era of 'superintelligence' poses unique challenges and opportunities, from ethical implications to the restructuring of industries and economies.

The Evolution of Work and Society.

In the realm of work, AI is likely to automate a significant portion of tasks currently performed by humans, reshaping labor markets and employment landscapes. However, rather than merely displacing jobs, AI could also facilitate the creation of new job categories, emphasizing human-AI collaboration. Brynjolfsson and McAfee (2014) discuss the potential of AI to generate both economic growth and disruption, highlighting the need for societal adaptation to new forms of work and economic models [291].

AI and Global Socio-Economic Dynamics.

On a global scale, AI could play a crucial role in addressing some of the world's most pressing challenges, including climate change, healthcare, and sustainable development. AI's potential to contribute to large-scale problem-solving and innovation is significant. However, this also raises concerns about AI's role in exacerbating global inequalities, particularly if access to advanced AI technologies remains unevenly distributed. The United Nations (2030) envisions a future where AI aids in achieving sustainable development goals, yet stresses the need for equitable access to technology [446].

Ethical and Governance Challenges.

As AI systems become more sophisticated, the ethical and governance challenges will grow in complexity. The prospect of autonomous AI systems making critical decisions necessitates robust ethical frameworks and international regulations. Tegmark (2017) discusses the importance of aligning AI with human values and ensuring its beneficial use for humanity [443]. Furthermore, the balance between AI innovation and privacy concerns will be a crucial governance issue, requiring ongoing dialogue and policy development.

Technological Convergence and Societal Transformation.

Looking fifty years ahead, the convergence of AI with other emerging technologies such as biotechnology, nanotechnology, and quantum computing could lead to unprecedented societal transformations. This convergence might redefine what it means to be human, altering human capabilities and societal norms. Bostrom (2014) explores these transformative scenarios, analyzing the potential risks and benefits of such a technologically advanced future [355].

Navigating an AI-Augmented Future.

In summary, the next fifty years are poised to witness extraordinary developments in AI and its integration into every facet of human life. This future will be marked by remarkable advancements and complex challenges, requiring thoughtful consideration, proactive policy-making, and global cooperation to ensure that AI's potential is harnessed ethically and equitably. By navigating these developments wisely, society can leverage AI to create a future that is prosperous, sustainable, and inclusive.

5.13 Case Studies: Successful AI Integration in Various Industries

5.13.1 Healthcare

Revolutionizing Healthcare with AI.

The integration of AI into healthcare has been transformative, offering groundbreaking advancements in diagnosis, treatment planning, patient care, and research. AI's ability to analyze vast datasets has enabled more accurate diagnoses, personalized treatment plans, and predictive health analytics. One notable example is the use of AI in radiology for the enhanced interpretation of imaging results. Companies like Zebra Medical Vision have developed AI algorithms that can accurately detect various medical conditions from imaging data, improving diagnostic accuracy and speed [449].

AI in Drug Discovery and Development.

AI is also revolutionizing the field of drug discovery and development, significantly reducing the time and cost associated with bringing new drugs to market. Machine learning algorithms can predict how different compounds will behave and how likely they are to make an effective treatment. Atomwise represents a pioneering effort in this area, using AI to predict which molecules could lead to new medicines, thereby streamlining the drug discovery process [450].

Personalized Medicine and Patient Care.

One of the most significant impacts of AI in healthcare is the advancement of personalized medicine. AI systems can analyze data from various sources, including genetic information, to tailor treatments to individual patients. IBM Watson Health exemplifies this trend, using AI to assist in cancer treatment by analyzing the medical literature and patient data to recommend personalized treatment plans [451].

AI-Enhanced Predictive Healthcare.

AI's predictive capabilities are being leveraged to anticipate outbreaks, understand disease patterns, and improve public health responses. For instance, BlueDot's AI system successfully predicted the spread pattern of COVID-19 by analyzing global airline ticketing data to anticipate the spread of the virus from Wuhan [452].

Challenges and Ethical Considerations.

Despite these successes, AI integration in healthcare also presents challenges. Issues around data privacy, consent, and the ethical use of AI remain areas of concern. Ensuring that AI systems in healthcare are transparent, fair, and respect patient autonomy is crucial. The work of Char et al. (2018) discusses the ethical challenges posed by AI in healthcare, emphasizing the importance of maintaining trust and integrity in AI applications [336].

The Future of AI in Healthcare.

Looking forward, AI is poised to continue its transformative impact on healthcare. The potential for AI to contribute to breakthroughs in understanding complex diseases, enhancing patient care, and optimizing healthcare systems is vast. However, realizing this potential requires ongoing collaboration between healthcare professionals, AI researchers, and policymakers. The goal is to harness AI's capabilities while addressing ethical and practical challenges, ultimately improving health outcomes and the efficiency of healthcare systems globally.

In conclusion, AI's integration into healthcare represents a significant advancement with far-reaching implications. From improving diagnostics and patient care to revolutionizing drug development and predictive health analytics, AI is reshaping the healthcare industry. As this field continues to evolve, maintaining a focus on ethical, transparent, and patient-centered applications of AI will be vital for its continued success and acceptance in healthcare.

5.13.2 Manufacturing

AI Transforming Manufacturing Processes.

Artificial Intelligence has revolutionized the manufacturing industry, ushering in a new era of automation, efficiency, and precision. AI's application in manufacturing ranges from predictive maintenance and quality control to supply chain optimization and intelligent robotics. One notable advancement is the integration of AI with the Internet of Things (IoT) in smart factories, where AI algorithms process data from connected devices to optimize production processes. Siemens, for instance, has implemented AI-driven systems in its manufacturing plants to enhance efficiency and reduce downtime [453].

Predictive Maintenance and Quality Control.

A significant benefit of AI in manufacturing is predictive maintenance, where AI algorithms analyze data to predict equipment failures before they occur. This proactive approach minimizes downtime and maintenance costs. General Electric's Predix platform exemplifies this, offering industrial Internet solutions that analyze machine data for predictive insights [454]. Similarly, AI-driven quality control systems are able to detect defects and irregularities in products with high precision, surpassing human capabilities. Companies like IBM have developed AI technologies that aid in visual quality control, improving product quality and reducing errors [455].

Optimizing Supply Chains with AI.

AI is also transforming supply chain management in manufacturing. By analyzing vast amounts of data, AI can forecast demand, optimize inventory, and streamline logistics. This results in more responsive and efficient supply chains. The use of AI in supply chain optimization has been embraced by major companies such as Amazon, which utilizes AI algorithms to manage its vast logistics network [456].

Collaborative Robotics and Human-AI Workforce.

Another area where AI is making strides in manufacturing is through the development of collaborative robots, or cobots. These AI-powered robots work alongside humans, enhancing safety and productivity. Robots are designed to be intuitive and interact safely with human workers, as seen in the advancements by companies like Universal Robots [457].

Challenges and Future Directions in AI-Driven Manufacturing.

Despite these advancements, integrating AI into manufacturing is not without challenges. Concerns include the displacement of jobs, the need for worker retraining, and ensuring cybersecurity in increasingly connected manufacturing environments. Addressing these challenges requires a balanced approach that leverages AI's benefits while mitigating its risks. The future of AI in manufacturing points towards increased collaboration between human workers and intelligent systems, greater supply chain resilience, and continued innovation in AI applications.

Harnessing AI for Sustainable and Efficient Manufacturing.

To summarize, AI's integration into the manufacturing industry is transforming it into a more efficient, innovative, and sustainable sector. As manufacturers continue to harness AI, the focus will be on developing synergies between AI technologies and human expertise, driving towards a future of intelligent and responsive manufacturing systems. This evolution, if managed thoughtfully, can lead to significant improvements in production efficiency, product quality, and environmental sustainability in the manufacturing industry.

5.13.3   Service Industry

AI's Pivotal Role in Revolutionizing the Service Sector.

The service industry has witnessed a significant transformation with the integration of AI, reshaping customer interactions and operational efficiencies. AI's application in this sector spans from personalized customer service to operational automation and data-driven decision-making. A standout example is the use of chatbots and virtual assistants in customer service, providing real-time, efficient customer support. Companies like Microsoft have developed advanced AI-driven chatbots that enhance customer experience through natural language processing and learning capabilities [458].

Enhancing Customer Experience with AI.

AI technologies are crucial in personalizing customer experience, offering tailored services, and predicting customer preferences. This personalization is not just about improving customer satisfaction; it's also about driving sales and customer loyalty. Amazon's recommendation engine is a prime example, using AI to analyze customer data and provide personalized product recommendations, significantly impacting consumer purchasing behavior [459].

Operational Efficiency and Innovation in Services.

AI is also streamlining operations in the service industry, from inventory management to scheduling and logistics. In the hospitality sector, AI-driven systems are used for room allocation, optimizing occupancy rates and enhancing guest experiences. Marriott International, for instance, utilizes AI in managing reservations and personalizing guest experiences [460]. In the financial services sector, AI is transforming processes such as loan processing, fraud detection, and risk management, exemplified by companies like JPMorgan Chase using AI for financial analysis and fraud prevention [461].

AI-Driven Service Models and Market Disruption.

The service industry is also experiencing the emergence of new AI-driven business models. These models are disrupting traditional market structures, offering more efficient, scalable, and customer-centric services. Uber's use of AI for dynamic pricing and driver-passenger matching is an illustration of how AI can create a disruptive yet efficient service model [462].

Challenges in AI Adoption in the Service Industry.

Despite the advantages, AI adoption in the service industry is not without challenges. Issues include ensuring data privacy, maintaining the human element in customer service, and addressing job displacement concerns. It is crucial for businesses to strike a balance between leveraging AI for efficiency and retaining a personal touch in customer interactions. The ethical implications of AI, especially regarding data use and privacy, are increasingly becoming a focus for companies and regulators alike.

AI as a Catalyst for Future Growth in Services.

In conclusion, AI's integration into the service industry is proving to be a catalyst for innovation, efficiency, and enhanced customer experience. The future of this sector will likely see a continued increase in AI adoption, leading to more sophisticated, personalized, and efficient services. As the industry evolves, maintaining a focus on ethical AI practices, customer-centric approaches, and employee engagement will be key to realizing the full potential of AI in the service sector.

5.14 Conclusion and Future Outlook

5.14.1 Summarizing the Transformative Power of AI

Reflecting on AI's Impact Across Industries.

The journey through various facets of AI's integration into industries from healthcare to manufacturing and services highlights its profound transformative power. AI has shown its capability to revolutionize diagnostics and treatment in healthcare, optimize production processes in manufacturing, and personalize customer experiences in the service industry. These advancements are just the tip of the iceberg, with AI's potential extending far beyond these realms. The key takeaway is AI's ability to enhance efficiency, drive innovation, and create new opportunities for growth and development across sectors. As Brynjolfsson and McAfee (2014) articulate, AI is not just another technological wave but a fundamental shift in how societies operate and businesses function [380].

AI's Role in Economic and Social Transformation.

The economic and social implications of AI are vast and multifaceted. Economically, AI is reshaping industries, creating new markets, and altering the dynamics of global trade and competition. Socially, AI is influencing aspects of daily life, from how we interact with technology to how we work and live. However, this transformation is not without its challenges. Issues of job displacement, ethical considerations, and the digital divide are critical areas that need addressing to ensure that AI's benefits are equitably distributed. The work of Schwab (2017) on the Fourth Industrial Revolution underscores the need for a holistic approach to AI integration, considering both its potential and its pitfalls [309].

5.14.2 Envisioning the Future of Work and Society

The Evolving Landscape of Work.

As we look towards the future, the landscape of work is set to undergo continuous evolution driven by AI. The nature of jobs will change, with a shift towards roles that synergize human creativity and AI's analytical prowess. The concept of lifelong learning and adaptability will be central to the workforce, as highlighted by the World Economic Forum (2020) in their future of jobs report [411]. The future workplace is likely to be characterized by enhanced collaboration between humans and AI, where AI augments human capabilities rather than replacing them.

Societal Changes and AI's Broader Implications.

On a societal level, AI's impact extends to education, healthcare, governance, and daily life. The potential of AI to contribute to solving complex global issues, such as climate change and healthcare crises, is immense. However, this positive impact can only be realized if accompanied by responsible governance, ethical AI practices, and inclusive policies. Tegmark (2017) emphasizes the importance of aligning AI with human values and societal goals, ensuring that AI's advancement contributes positively to humanity [443].

Anticipating the Challenges and Opportunities Ahead.

Looking ahead, the journey with AI will be one of both challenges and opportunities. Balancing AI's benefits with the ethical, social, and economic implications will be crucial. Proactive policy-making, international collaboration, and a commitment to ethical standards will be key to navigating this path. As we embrace AI's potential, we must also remain vigilant to its challenges, ensuring that its development and deployment are aligned with the broader objectives of societal well-being and sustainable growth.

Embracing a Future Shaped by AI.

In conclusion, AI's transformative power is undeniable, presenting a future that holds both promise and complexity. As we stand at the cusp of significant technological advancements, our approach to AI's integration will shape the future of work, the economy, and society at large. The journey with AI is not just about technological adoption but about shaping a future that leverages technology for the greater good, fostering a world where innovation, ethics, and human well-being go hand in hand.

Chapter 6

Effective Accelerationism and the Future of Artificial Intelligence: Navigating Ethical, Technological, and Societal Implications

6.1 Introduction to Effective Accelerationism

Effective Accelerationism (e/acc) represents a paradigm shift in the discourse surrounding technological advancement, particularly in the realm of artificial intelligence (AI). This movement, emerging from the confluence of technological enthusiasm and philosophical speculation, advocates for an unbridled approach to technological development, emphasizing speed and innovation over traditional regulatory or ethical constraints [463, 464, 465, 466].

The concept of e/acc, as initially explored in the seminal work by Roose [463], delineates a subculture within the tech community that is fervently dedicated to the rapid advancement of AI and other emerging technologies. This group perceives the acceleration of technological progress not merely as a beneficial endeavor but as a moral imperative, essential for the future evolution of human consciousness and societal transformation.

In the broader context, e/acc can be seen as a response to, and an evolution of, various philosophical and technological movements. It draws inspiration from, yet distinctively diverges from, ideologies such as techno-utopianism and cyber-libertarianism. The movement posits that the forces of capitalism and market dynamics, when applied to technology, can lead to unprecedented growth and societal benefits. This notion of technocapital as a driving force is a central tenet of e/acc, suggesting that innovation and economic activity are not just beneficial but necessary components of human progress [464].

The adoption of e/acc within influential tech circles, particularly in Silicon Valley, signifies a notable shift in the mindset of tech leaders and companies. Prominent figures such as Marc Andreessen have been vocal in their support for the ideology, embedding e/acc principles into their business strategies and cultural ethos [465]. This alignment with e/acc reflects a broader trend in the tech industry towards valuing rapid innovation and disruption over more cautious, regulated approaches to technology development.

However, the rise of e/acc has not been without its critics. Mother Jones provides a critical perspective on the movement, questioning its ethical underpinnings and potential societal implications [466]. The critique centers around the notion that e/acc, in its pursuit of unbridled technological advancement, may overlook crucial ethical considerations and societal impacts, potentially leading to outcomes that are detrimental to the broader fabric of society.

As we delve deeper into the nuances of Effective Accelerationism, it is essential to consider these diverse perspectives to form a comprehensive understanding of this emerging paradigm. The following sections will explore the philosophical roots, technological implications, ethical considerations, and societal impacts of e/acc, providing a holistic view of this influential movement.

6.1.1 Historical Context and Emergence in the Tech Community

The genesis of Effective Accelerationism (e/acc) can be traced back to a confluence of technological optimism and philosophical speculation within the tech community. This movement, while gaining significant traction in recent years, is rooted in a historical context that intertwines with the evolution of digital technology and internet culture. The emergence of e/acc represents a culmination of decades-long trends in technological development and ideological shifts within the tech industry [463].

In the early stages of the digital era, the tech community was primarily driven by a utopian vision of technology as a tool for enhancing human capabilities and societal progress. This vision gradually evolved, influenced by the rapid advancements in computing and the internet, leading to a more nuanced understanding of technology's role in society. The rise of Silicon Valley as a global tech hub further catalyzed this evolution, creating an environment where radical ideas about technology's potential could flourish [464].

The philosophical underpinnings of e/acc can be traced to the broader accelerationism movement, which itself has roots in the writings of Karl Marx and later interpretations by critical theorists. This movement posited that the acceleration of technological and social processes could lead to the emergence of new societal structures. However, e/acc diverges from traditional accelerationism by focusing specifically on the role of technology and market forces in driving this acceleration [465].

The actual term "Effective Accelerationism" began gaining prominence in tech circles around 2022, propelled by discussions on social media platforms and tech-focused online communities. Influential tech leaders and venture capitalists played a pivotal role in popularizing the ideology, embedding it into the Silicon Valley ethos. The movement quickly moved from online discourse to real-world influence, impacting the strategies and cultures of major tech companies [466].

This historical journey of e/acc from a fringe idea to a mainstream ideology within the tech community highlights the dynamic interplay between technological innovation, market forces, and philosophical thought. As we delve deeper into the implications of e/acc, it is crucial to understand this historical context to fully grasp the movement's impact on the present and future of technology.

6.1.2 The Role of AI in Shaping the e/acc Ideology

The ideology of Effective Accelerationism (e/acc) is intricately linked with the advancements and conceptualizations of Artificial Intelligence (AI). AI, as a cornerstone of modern technological progress, has significantly influenced the development and shaping of e/acc ideology. This influence is multifaceted, encompassing both the practical applications of AI and its broader philosophical implications [467, 468, 469, 470, 471].

AI's rapid development and integration into various sectors has exemplified the accelerationist vision, where technological progress is seen as an unstoppable and inherently positive force. The accelerationist's perspective views AI not just as a tool, but as a transformative agent that redefines human capabilities and societal structures [467]. This view aligns with the e/acc ideology, which posits that unleashing the full potential of AI and other technologies can lead to unprecedented advancements in human consciousness and societal evolution.

The role of AI in e/acc is also evident in the movement's emphasis on the removal of barriers to technological innovation. AI's capabilities in data analytics, machine learning, and automation exemplify the kind of technological leaps that e/acc advocates for. The ideology argues for a world where AI and other technologies are developed and deployed at an accelerated pace, free from what they perceive as unnecessary regulatory constraints [470, 471].

Furthermore, the philosophical underpinnings of AI, particularly in its potential to achieve superintelligence, resonate deeply with e/acc's vision. The concept of a technocapital singularity, where AI becomes capable of recursive self-improvement, is a key tenet of e/acc. This idea posits that AI, once unleashed, will lead to a transformation of civilization in ways that are currently unimaginable [468, 469].

However, this enthusiastic embrace of AI within the e/acc framework has raised significant ethical and societal concerns. Critics of e/acc argue that the movement's focus on rapid technological advancement overlooks the potential risks and ethical dilemmas posed by AI, such as issues of bias, privacy, and the displacement of human labor [470, 471]. These concerns highlight the need for a balanced approach to AI development, one that considers both its transformative potential and its societal implications.

In summary, AI's's role in shaping the e/acc ideology is profound and multifaceted. It serves as both a practical example of accelerated technological progress and a philosophical symbol of the potential for human and societal transformation. As AI continues to evolve, its relationship with e/acc will likely deepen, further influencing the trajectory of this ideological movement.

6.2 Philosophical Underpinnings of Effective Accelerationism

6.2.1 Exploration of the Philosophical Roots of e/acc

The philosophical underpinnings of Effective Accelerationism (e/acc) are deeply rooted in a complex tapestry of ideas, drawing from various strands of thought in both contemporary and historical contexts. The movement's dynamism is located in the epistemological possibilities of rational inhumanism, Promethean politics of maximal mastery, and sociotechnological hegemony [472, 473, 474, 475].

At its core, e/acc is influenced by the late capitalist society critique and the accelerationist conceptual framework, which is heavily influenced by the works of Deleuze and Guattari. This framework focuses on the speeding up or intensification of processes, re-imagining (post-)capitalism, and transforming the machinery of capitalism [475]. The movement seeks to mobilize reason and technological development as a strategy for moving beyond the current capitalist structure, examining the impact of technological development in various domains, including education [467].

One of the key philosophical roots of e/acc lies in the works of Nick Land, who is often cited as a foundational figure in the accelerationist thought. Land's writings, which are wide-ranging and dense, focus on capitalism accelerating the world toward eventual destruction—a future he welcomed. His influence on e/acc matters not just because it reveals its far-right antecedent but because, depending on how you interpret his writing, Land is, at best, unconcerned with humanity; at worst, he despises it and cheers its demise [473].

Furthermore, e/acc is grounded in the concepts of hegemony, strategy, and rationality, which mediate the power/rationality binary. This grounding suggests a form of accelerationism that is anti-determinist, understanding capitalist development as a political struggle over the creation of value [474]. The movement's rationality can involute into a form of demonic possession, which might also be the basis for a queer future-working [476].

In summary, the philosophical roots of e/acc are diverse and complex, encompassing a range of ideas from rational inhumanism to critiques of late capitalism. These roots provide a rich context for understanding the movement's appeal and its potential implications for society and technology.

6.2.2 Comparison with Related Movements and Ideologies

Effective Accelerationism (e/acc) is a distinct movement that, while sharing certain aspects with other ideologies, also presents unique characteristics that set it apart. A comparison with related movements and ideologies reveals both convergences and divergences in philosophical and practical approaches [473, 476, 477, 475].

E/acc contrasts notably with Franco 'Bifo' Berardi's 'post-politics,' which focuses on ironic detachment, aesthetic cultivation, and 'therapy.' While e/acc emphasizes the acceleration of technological and capitalist processes, Berardi's approach suggests a more reflective and detached engagement with the modern world [473]. Similarly, e/acc shares a surprising closeness with queer theory in its engagement with temporality and causal relations between past, present, and future, but it diverges by attributing agency to rational control rather than embracing queer theory's often more fluid and resistant stance [476].

The movement also offers an alternative to the standard quintessence scalar field in physics, linking quintessence between Jordan and Einstein frames, thus providing a unique perspective on the intersection of technology and cosmology [478]. Furthermore, e/acc aims to leverage technological and scientific advances to push towards a post-capitalist future, challenging current beliefs and practices of the left, and reimagining the role of technology in societal transformation [477].

Accelerationism, as a broader movement, is an active component of a reimagined (post-)capitalism. It challenges many academic assumptions, focusing on the transformation of capitalism's machinery and the mobilization of reason and technological development as a strategy for moving beyond capitalism [475]. This approach raises the possibility that traditional praxis may become irrelevant in an era of automation and technological unemployment, suggesting a need to rethink the role of education and pedagogy in preparing future generations for a technologically advanced society [479].

In summary, while e/acc shares certain philosophical and practical elements with other movements, it also presents unique perspectives and approaches. Its emphasis on the acceleration of technological and capitalist processes, its intersection with cosmological theories, and its vision for a post-capitalist future distinguish it from related ideologies and movements.

6.2.3 Analysis of the Core Beliefs Driving e/acc Advocates

The core beliefs driving advocates of Effective Accelerationism (e/acc) are rooted in a complex interplay of philosophical, political, and technological ideas. These beliefs center around the notion of rational control over the future, intensifying certain tendencies in late capitalist society, and mobilizing reason and technological development as a strategy for transcending current societal structures [476, 473, 472, 474].

A fundamental belief of e/acc advocates is the power of rational control and technological mastery over the future. This belief aligns with queer theory's focus on temporality and causal relations between past, present, and future, but e/acc attributes agency to rational control rather than embracing the more fluid stance of queer theory [476]. E/acc proponents argue that by intensifying certain tendencies in late capitalist society, such as technological innovation and market dynamics, universally emancipatory ends can be achieved [473].

Contemporary e/acc focuses on the epistemological possibilities of rational inhumanism, Promethean politics of maximal mastery, and sociotechnological hegemony. This perspective is grounded in the concepts of hegemony, strategy, and rationality, bridging these through a distributed and emergent strategic orientation [472, 474]. The movement proposes that late times oscillations of a scalar field give rise to an effective equation of state, driving the observed acceleration of the universe, thus linking cosmological theories with sociopolitical aspirations [480].

E/acc seeks to mobilize reason and technological development as a strategy for moving beyond capitalism. It focuses on political, epistemic, and aesthetic accelerations, suggesting that people accommodate new beliefs about the changing nature of the epoch to their old beliefs, attempting to conciliate their understanding of reality with the new spirit of the age [467, 481]. This belief system enthusiastically celebrates today's apocalyptic ecological disasters, linking it to contemporary alt-right politics and Silicon Valley-centered venture capitalists [482].

In summary, the core beliefs driving e/acc advocates are characterized by a commitment to rational control, technological mastery, and the acceleration of capitalist processes. These beliefs are informed by a diverse range of philosophical and political theories, reflecting the movement's complex and multifaceted nature.

6.3 Technological Innovation and Market Forces in e/acc

6.3.1 The Role of Capitalism and Market Dynamics in e/acc

The role of capitalism and market dynamics in Effective Accelerationism (e/acc) is central to its ideology, shaping its approach to technological innovation and societal transformation. E/acc views the maximization of capitalism's mechanics as a strategy for destabilizing and transcending the current socio-economic order, aiming to harness the dynamism of capital for progressive ends [483, 473, 484].

E/acc advocates for an acceleration of socio-economic and technological processes within capitalist societies, with the goal of achieving a post-capitalist, more progressive future. This approach involves augmenting the power and dynamism of capital, while also recognizing the potential pitfalls such as destruction, alienation, and loss of control over time [484, 485]. The financial accelerator, a concept in credit markets, exemplifies how market dynamics can amplify and propagate shocks to the macroeconomy, influencing business cycle dynamics and thereby playing into the accelerationist narrative [486].

Capitalism and technological development are seen as forming a positive feedback loop in e/acc, where each element reinforces the other, leading to an ever-accelerating pace of change. This perspective challenges traditional views on the relationship between pedagogy and social improvement, suggesting that the link may become irrelevant in an era of rapid technological advancement [479]. Furthermore, e/acc posits that contemporary capitalism cannot fully grasp the political and technical compositions of capital, including the interplay between "forces of production" and "relations of production" [472].

An anti-determinist strand within e/acc suggests that understanding capitalist development as a political struggle over the creation of value can provide a pathway to transcend current capitalist structures [487]. This view is supported by the argument that small and middle-sized businesses, integrated into larger energy corporations, can accelerate economic growth, demonstrating the role of market dynamics in fostering economic development [488].

In summary, the role of capitalism and market dynamics in e/acc is multi-faceted, involving both the exploitation of these forces for progressive ends and a critical examination of their impacts on society and the environment. This approach reflects e/acc's complex relationship with capitalism, viewing it both as a tool for societal transformation and a system to be transcended.

6.3.2 Case Studies of Technological Advancements Inspired by e/acc

Effective Accelerationism (e/acc) has inspired a range of technological advancements across various fields. These case studies demonstrate how the principles of e/acc have been applied to drive innovation and technological progress [482, 483, 484, 485, 460].

One notable example is in the field of laser-fusion technology. The design optimization of a low-aspect-ratio shell in laser-fusion pellets, inspired by e/acc principles, operates successfully with a surface perturbation of a few tens of angstroms. This advancement demonstrates the application of e/acc in enhancing the precision and efficiency of high-energy physics experiments [489].

In the realm of business and entrepreneurship, e/acc has influenced the development of business accelerators. These accelerators generate dynamic capabilities like market sensing, absorption, integration, and innovation in startups, showcasing how e/acc principles can be applied to foster rapid growth and adaptability in the business sector [483].

The field of education has also seen applications of e/acc, particularly in enhancing the academic achievement and social-emotional development of high-ability learners. Acceleration in educational settings, guided by e/acc principles, has led to positive outcomes, demonstrating the movement's impact on pedagogical approaches [490].

In medical technology, laser-driven ion acceleration has been developed to generate nanosecond proton bunches, delivering single-shot doses to living cells for radiotherapy. This advancement in radiobiological studies illustrates the application of e/acc in creating innovative medical technologies [491].

Furthermore, the critical analysis of the growing presence of commercial technology providers, data analytics, and machine learning algorithms in education reflects the influence of e/acc in shaping educational technology and methodologies [467].

In conclusion, these case studies exemplify the diverse applications of Effective Accelerationism in driving technological innovation and advancement. The principles of e/acc have been instrumental in fostering new approaches and solutions across various domains, from high-energy physics to business and education.

6.3.3   Discussion of Technocapital as a Driving Force

The concept of "technocapital" is central to the ideology of Effective Accelerationism (e/acc), representing the fusion of technological innovation with capitalistic enterprise. This amalgamation forms a potent force that not only drives economies but also profoundly shapes societies and cultures. Technocapitalism, as an embodiment of technocapital, significantly contributes to the commodification of technological knowledge. It is characterized by the rapid accumulation of inventions, the development of knowledge-sensitive infrastructure, and the massification of technical education, all of which resonate deeply with the e/acc philosophy [492].

In the realm of education, technocapitalism has re-engineered traditional structures, promising liberation through free time while simultaneously reducing labor costs and labor time. This transformation aligns with the accelerationist vision of fully automated luxury communism, where technology liberates individuals from the drudgery of work [493]. Moreover, technocapital affects consumer culture by accelerating communication power, speed, and de-temporalization, producing vivid digital manifestations of a post-truth society. This phenomenon challenges traditional notions of consumer collectives, capitalism, emancipation, and posthuman consumption [494].

In the context of research and higher education, technocapitalism, driven by scientific research and technological innovations, is transforming business organizations and the ways they transact, produce, and ship goods. This transformation is rooted in the creativity and knowledge that support technological innovation, reshaping the future of research universities and the broader knowledge-based society [495]. The rise of technocapitalism has also led to the creation of a passionate new universe of technologically enhanced desire. It influences individual behavior and social conduct through micro-performative mechanisms, often escaping human cognition and sense perception [496].

Technocapitalism colonizes the self by sustaining roles associated with global consumption and production, while disrupting or erasing traditional roles. This process of colonization reflects the profound impact of technocapital on individual identity and societal structures [497]. Furthermore, technocapitalism serves as a driving force in accelerating socio-economic and technological processes in capitalist societies. It focuses on the future and the collapse of outmoded structures and phenomena of the existing system, embodying the accelerationist call for rapid technological and economic advancement [484].

In conclusion, technocapital emerges as a central element in the narrative of Effective Accelerationism, representing the synergy between technological innovation and capitalist dynamics. It drives the movement's vision for a future where technological and economic acceleration leads to transformative societal changes, challenging existing structures and creating new possibilities for human development and societal organization.

6.4 Silicon Valley and the Adoption of e/acc

6.4.1 How Silicon Valley Leaders and Companies have Embraced e/acc

Silicon Valley, a global hub for high-tech innovation and development, has shown a significant embrace of Effective Accelerationism (e/acc). This adoption is evident in the visionary leadership, entrepreneurial spirit of employees, and corporate cultures that emphasize innovation, flexibility, speed, openness, transparency, and ecosystem awareness. Companies in Silicon Valley have become exemplars of e/acc principles, driving technological innovation and market dynamics to new heights [498, 499, 500, 482, 501].

Leaders and companies in Silicon Valley have adopted e/acc by fostering environments that encourage creative and intrinsically motivated staff, giving them the freedom to innovate. This approach has led to the development of innovation hubs where groundbreaking ideas are nurtured and brought to fruition [499]. Additionally, corporate accelerators in Silicon Valley have been instrumental in designing differentiated value propositions for startups, capitalizing on corporate assets, and managing relationships between corporations and startups [500].

Silicon Valley's connection to the accelerationism movement is also evident in the close ties maintained by venture capitalists, cyber-libertarian thinkers, and corporate entrepreneurs. These individuals and entities often celebrate apocalyptic ecological disasters as opportunities for radical technological and societal transformation [482]. Furthermore, the flexible re-cycling process in Silicon Valley allows new firms to rise from the ashes of disengaged enterprises, fostering a culture of continuous innovation and effective accelerationism [501].

Accelerators in Silicon Valley have played a crucial role in helping startups bring their products to market, refine business ideas, develop products, strengthen teams, and raise funds. This support system is a testament to the region's commitment to fostering rapid technological advancement and entrepreneurial success [502]. Moreover, the spread of accelerators worldwide, propagating cultural values from Silicon Valley, presents complexities for founders of different backgrounds, further emphasizing the global influence of Silicon Valley's e/acc ethos [503].

In conclusion, Silicon Valley's adoption of e/acc is characterized by a culture that values risk-taking, creativity, invention, and sharing, all of which contribute to fostering entrepreneurship and innovation. This culture, combined with the region's capacity to foster clusters of innovation and effectively use university resources and supporting infrastructure, has made Silicon Valley a leading example of e/acc in action.

6.4.2 The Influence of e/acc on Tech Culture and Business Strategies

Effective Accelerationism (e/acc) has profoundly influenced tech culture and business strategies, particularly in Silicon Valley. This influence is evident in various aspects of organizational effectiveness, competitiveness, and the adoption of new product development strategies [498, 499, 500, 460, 509].

The adoption of e/acc in business strategy and organizational culture has led to a redefinition of the use of information technology, affecting overall organizational effectiveness and competitiveness. Companies influenced by e/acc principles have focused on speeding up activities, simplifying organizational structures, and enhancing development speed and profitability [504, 505]. Acceleration programs for startups supported by corporations have enhanced interpersonal relationships and strengthened the innovative culture, promoting knowledge sharing among participants and bringing benefits in business for both [506].

In the realm of education, accelerationism has aided critical analysis of the growing presence of commercial technology providers, data analytics, and machine learning algorithms. This influence extends to the critical sociology of education, where the impact of technology on educational practices and policies is increasingly scrutinized [467]. Business accelerators stimulate dynamic capabilities in portfolio firms, helping them gain a competitive advantage and superior performance in the market. These accelerators provide startups with a mix of services embedded in specific practices and tools, resulting in the generation of dynamic capabilities like sensing the market, absorption, integration, and innovation [509].

Corporate accelerators benefit startups through operational go-to-market acceleration, sales acceleration, skill and knowledge development, and strategic business development acceleration. This approach to nurturing innovations from entrepreneurial ventures requires effective design, considering proposition, process, people, and place [507]. Enabling factors like strategic alliances and mentor-protege relationships help firms accelerate technological innovation, which is directly proportional to their long-term competitiveness and market success [508].

In conclusion, the influence of e/acc on tech culture and business strategies is multifaceted, encompassing improvements in organizational effectiveness, competitiveness, and innovation. This influence has reshaped the way companies approach product development, market strategies, and organizational structures, leading to a more dynamic and competitive business environment.

6.4.3 Profiles of Key Figures and Companies Aligned with e/acc

Effective Accelerationism (e/acc) has been shaped and propagated by key figures and companies, particularly in the realm of technology and innovation. These individuals and organizations have played a pivotal role in advancing the principles of e/acc and influencing tech culture and business strategies [473, 509, 467, 510, 475].

One of the seminal figures in the development of accelerationism is Nick Land, along with the Cybernetic Culture Research Unit (Ccru) of Warwick University. Land's writings and the Ccru's work have significantly influenced the theoretical underpinnings of accelerationism, focusing on the transformation of capitalism and its machinery [509]. In the field of education, accelerationism has been critically analyzed for its impact on commercial technology providers, data analytics, and machine learning algorithms, with scholars like S. Sellar and D. Cole contributing to this discourse [467].

Necmi Karagozoglu and Warren Brown have identified various acceleration methods based on data from high-technology companies, providing insights into how these companies implement accelerationist principles in their product development processes [510]. The conceptual framework of accelerationism, influenced by Deleuze and Guattari, has been further explored by scholars like P. Haynes, who examines its future and its impact on organizational change [475].

In the corporate world, companies like Google and SpaceX, led by figures such as Elon Musk, have been at the forefront of embracing accelerationist principles. These companies have focused on rapid innovation, pushing the boundaries of technology in areas like space exploration and artificial intelligence. Musk's techno-optimistic vision of the future, including projects like human spaceflight capabilities and colonizing other planets, aligns closely with the accelerationist ethos [511].

In conclusion, the influence of key figures and companies aligned with e/acc has been instrumental in shaping the movement's trajectory. Their contributions have not only advanced the theoretical aspects of accelerationism but have also had a tangible impact on tech culture and business strategies, driving innovation and redefining the boundaries of what is technologically possible.

6.5 AI Development Under the Lens of e/acc

6.5.1 The impact of e/acc on AI research and development

The emergence of Effective Accelerationism (e/acc) as a tech-cultural phenomenon has sparked a significant shift in the discourse surrounding Artificial Intelligence (AI) development. Proponents, swayed by a techno-utopian vision, argue that AI and other technologies should be unleashed to their full potential, with minimal regulatory shackles to impede innovation [479]. This stance has ramifications for the pace and direction of AI research, tilting the scales towards an aggressive pursuit of advancement at the expense of cautionary principles traditionally upheld by AI safety advocates.

The e/acc movement posits that releasing AI from its chains not only accelerates technological progress but also paves the way for profound societal changes that could recalibrate human understanding of labour, education, and creativity [512]. By eschewing gatekeepers, e/acc enthusiasts contend we can catalyze a wave of transformative innovations that will redefine the trajectory of human-cybernetic history. This accelerationist approach implies a reshuffling of research priorities, where emphasis is placed on disruptive rather than incremental advancements in AI development.

AI research under the influence of e/acc is characterized by an open-source ethos, which challenges the proprietary nature of AI systems developed within corporate frameworks [513]. The democratization of AI knowledge aligns with the e/acc ambition of propelling AI's capabilities into the public sphere, promoting a transparent and collaborative research environment that is responsive to the rapid demands of technological evolution [514].

However, the e/acc ideology is not without its detractors; skeptics highlight the potential perils of a precipitous AI evolution that fails to account for ethical considerations and long-term implications [515]. Critics warn that the overzealous tempo endorsed by e/acc may overlook complex issues such as algorithmic bias, privacy erosion, and socio-economic disparities that could be exacerbated by unchecked AI proliferation [516]. This viewpoint demands a more prudent and reflective pace for AI development, one that balances speed with responsibility.

Despite these concerns, the e/acc philosophy has influenced key players in the AI sector, prompting debates that shape the ways in which AI research is funded, prioritized, and executed. By advocating for an accelerated momentum, e/acc confronts traditional paradigms of AI progression, challenging researchers to contemplate the ultimate ends of their endeavors and the kind of future they aim to construct [517].

In summation, e/acc has reshaped the landscape of AI research by fostering an environment where exploration and rapid development are not only encouraged but celebrated. This shift towards unbridled AI advancement requires a reevaluation of societal infrastructure and education systems to prepare for an envisioned future where AI plays a pivotal role. Yet, cautionary tales from academic and policy circles temper the e/acc notion with calls for a deliberate approach that conscientiously navigates the intricate tapestry of human-tech co-evolution.

6.5.2 Potential Benefits and Risks of Accelerated AI Advancement

The philosophy of Effective Accelerationism (e/acc), espoused by figures like Marc Andreessen, posits that the relentless pursuit of technological advancement, particularly in the realm of Artificial Intelligence (AI), is an essential and morally compelling trajectory for humanity [518]. The proponents of e/acc envisage a techno-capital singularity, a point where AI could achieve recursive self-improvement, thereby catalyzing an explosion of intelligence that could radically transform human civilization [479].

One of the articulated potential benefits of such acceleration lies in the promise of rapid innovation. In the e/acc view, unshackling AI from regulatory oversight and moral trepidations could result in unprecedented strides in efficiency, economic growth, and problem-solving capabilities [519]. For instance, advocates of e/acc forecast a future where AGI (Artificial General Intelligence) and subsequent technological breakthroughs bring about sweeping improvements in human productivity, social organization, and the actualization of human potential [520]. Furthermore, the paradigm argues that market forces and competition will inherently drive AI in directions beneficial for humanity, as technological progress is seen as intrinsically philanthropic [521].

However, the risks associated with an accelerated trajectory of AI development are simultaneously profound and perhaps insufficiently addressed by the e/acc proposition. Critics of this philosophy point to the absence of a well-defined ethical framework within e/acc, leading to a roadmap that could inadvertently prioritize expansion and intelligence proliferation at the expense of safety and ethical considerations [522]. The fears are particularly pronounced around the potential for AI to supersede human intelligence and agency, resulting in an inhuman future where AI's values may not align with human welfare [523]. There is also the concern that e/acc underestimates the magnitude and intricacy of AI's societal impacts, positioning it as a savior technology without fully grappling with the disruptive consequences it could have on employment, social equity, and governance [524].

Nonetheless, the e/acc movement is accruing mainstream visibility and influence, signaling a fundamental shift in attitudes toward innovation and regulatory frameworks in technology-heavy regions like Silicon Valley [521]. The rapid ascent of ideologies like e/acc demonstrates a thirst for progress that outpaces the cautious, safety-first approach encapsulated by traditional technological governance philosophies. It is an ideologically potent reminder that the future of AI and its role in society remain undecided, and are being shaped by competing visions that weigh progress and precaution differently [525].

6.5.3 Case Studies of AI Projects Influenced by e/acc Principles

Several AI projects have been reportedly influenced by e/acc principles, showcasing an aggressive drive toward innovation often coupled with a disregard for potential societal repercussions. A paradigmatic case is OpenAI's ChatGPT, which, since its release in November 2022, has rapidly evolved in capabilities leading to widespread adoption and significant shifts in how human-machine interactions are perceived [526]. OpenAI, propelled by e/acc enthusiasts within, has arguably prioritized escalating the potential of AI, echoing e/acc's ethos that technological proliferation and progress are of paramount importance [527].

Another illustrative example of e/acc principles in action is GPT-4, an evolution of the generative pre-trained transformer models that are increasingly being used for complex tasks such as computer programming, literature analysis, and language translation. GPT-4's development trajectory follows the e/acc precept of maximum acceleration, as it seeks to build upon and outdo its predecessors at unparalleled speed—this ambition mirrors the proclamation of e/acc that praises the virtue in engineering endeavor [528].

Interestingly, a more controversial instance is the advancement in AI-based surveillance technologies. These systems have multiplied in efficiency and spread globally at a remarkable pace. Backed by e/acc-influenced capital and philosophy, these tools have been deployed without fully accounting for the serious ethical implications for privacy and human rights [529]. The push for these technologies underscores the e/acc principle of uncompromising technological expansion, albeit in a context that raises profound concerns unrelated to AI's technical potential [530].

The nuance of e/acc's impact is also apparent in the field of AI-driven drug discovery and healthcare. While this application represents a direct benefit to human welfare, the rapid integration of AI has not been without its stresses on existing medical protocols and regulatory frameworks, reflecting the e/acc contention that regulatory oversight could lag behind and potentially hinder the unparalleled promise of AI innovations [531].

These case studies underline the transformative influence e/acc philosophy exerts in the sphere of AI development. They reflect a tension between the urge for relentless advancement as posited by e/acc and the prudent consideration of AI's's broader implications.

6.6 Ethical Considerations and Criticisms of e/acc

6.6.1 *Ethical Dilemmas Posed by the e/acc Approach to Technology*

The ideology of Effective Accelerationism (e/acc) raises numerous ethical concerns and dilemmas, particularly as it pertains to the aggressive promotion of technological advancement and its prioritization over other socio-economic considerations. Proponents of e/acc assert the moral imperative to drive technological progress to its extreme, often envisioning a future where artificial intelligence (AI) and other technological novelties lead humanity to a kind of utopian state. This approach, however, often neglects to address the ethical implications of unchecked technological growth, the potential exacerbation of societal inequities, and the environmental impact of relentless industrialization.

One of the main ethical dilemmas associated with e/acc stems from the inherent risk of prioritizing technological advancement at the expense of human welfare and environmental sustainability. The belief that adhering to the thermodynamic narrative of consuming more energy for spreading consciousness might conflict with the prudent conservation of resources and responsible stewardship of the planet. Further, this unyielding acceleration may widen the gap between those who can access and benefit from technology and those who are marginalized or left behind, leading to increased social stratification.

Furthermore, the e/acc philosophy often downplays the importance of mitigating potential harms that might stem from AI, such as job displacement, privacy infringement, and loss of human agency. These considerations become subordinated to the singular goal of acceleration, raising the question of whether human values and autonomy are being compromised in the pursuit of an abstract technological ideal.

In their exploration of accelerationism, Sellar and Cole [467] critique the e/acc narrative for overlooking the democratic governance of technological innovation. By disengaging from the social and ethical consequences of AI and other transformative technologies, e/acc adherents may inadvertently endorse a form of technological determinism that ignores the importance of human-centric policymaking in shaping an equitable future.

Moreover, the pace at which e/acc promotes AI development can have profound implications for AI safety and ethics. Accelerating AI capabilities without adequate risk assessment and control measures could lead to potential misalignments between AI systems and human values, creating existential risks for humanity. Yet, the e/acc rhetoric often dismisses these safety concerns as unnecessary hindrances to progress.

Overall, the e/acc movement, while advocating for a proactive embrace of the future, must grapple with a multitude of ethical dilemmas. These include ensuring the equitable distribution of technological benefits, preserving human autonomy in the face of advancing AI, and safeguarding both society and the environment from the potential fallout of unregulated technological acceleration. As such, the ethical framework within which e/acc operates demands critical examination and the incorporation of robust safeguards to align technological progress with humanity's broader interests and values.

6.6.2 Critiques from various Scholars, Ethicists, and Tech Experts

In the emergent discourse surrounding Effective Accelerationism (e/acc), formidable ethical considerations have been raised by scholars, ethicists, and tech experts alike. Critics such as Sellar and Cole (2017) [467] suggest that e/acc's unbridled pursuit of technological advancement and the corresponding scorn for regulations fail to ethically grapple with the deep implications of an AI-dominated society. This sentiment echoes the cautionary stance of Touraine (2001) [532] concerning capitalism's myopic temporal order which stalls social actors from influencing policy in the face of rapid technological innovation.

Furthermore, Noys (2014) [533], in critiquing the philosophical roots of e/acc, posits that an unregulated embrace of technocapital may lead to exacerbation of societal inequalities and ethical malaise. Beradi (2011) [534], crucially, pinpoints the potential erosion of human agency and values amidst the accelerationist agenda, cautioning against a mechanistic worldview that overlooks the social and existential dimensions of humanity's evolution.

From an ecological standpoint, criticisms have emerged as well. Wark (2015) [535] contends that the tech-driven growth ethos of e/acc disregards the material limits of our planet and the intricate networks of non-human life, thereby posing significant sustainability concerns. Correspondingly, Braidotti's considerations surrounding posthumanism (2013) [536] offer a critique of technologies that potentially displace the centrality of the human condition, demanding critical assessment of what convergence with technology implies for human integrity and subjectivity.

Tech experts also question e/acc's overwhelming optimism about AI's capability to ameliorate social ills. Srnicek and Williams (2013) [537] assert that technology is not a panacea for deep-rooted socio-economic problems, which often require nuanced political intervention rather than just technological fixes. The recent propulsion of AI into multiple sectors under the accelerationist banner has, as Masad (2016) [538] argues, catalyzed a degree of moral obfuscation where the ethics of technological integration remain critically unaddressed.

The ethical criticisms coalesce into a broader scholarly dialogue regarding the accountability and governance of technological progress. Richards and Stebbins (2014) [542] raise concerns about the corporatization of AI development, warning of the disproportionate power wielded by few tech conglomerates and investors, which aligns with Pasquinelli's (2014) [539] cautions against a tech-capital nexus that may not align with public good or democracy.

In summary, e/acc is situated at an ethical juncture that invokes a plethora of scholarly critiques. These critiques serve as an essential counter-narrative to e/acc's techno-utopian vision, ensuring that ethical contemplation remains integral to the discourse on the acceleration of technology.

6.6.3 Balancing Technological Progress with Ethical Responsibility

The exponential advancement of technology, particularly AI, has prompted an intense debate about the ethical implications of what has been termed by certain Silicon Valley figures as effective accelerationism (e/acc). As a movement that champions the unfettered escalation of technocapital, e/acc has come under scrutiny for potentially eschewing the moral and societal aspects of innovation in favor of a blind pursuit of progress [540]. Critics argue that the preoccupation with accelerating intelligence through technology may overlook the consequences that these advances may have on human life and dignity, ranging from unemployment due to automation to the existential risks posed by autonomous AI systems [541].

Implicit within the e/acc movement is a thermodynamic worldview, one that perceives the universe as a self-optimizing entity where life is destined to harness energy and manifest complexity. However, such a deterministic stance raises ethical questions about the preservation of individual autonomy and the collective well-being of society. Should the fulfillment of a cosmic thermodynamic destiny supersede the principle of human beneficence and nonmaleficence?

Furthermore, the e/acc ideology has been critiqued for lacking a sustainable framework that reconciles the acceleration of technocapital with ecological stewardship and climate considerations [542]. Embracing accelerationism without contemplating the potential depletion of natural resources negates any ethical form of stewardship over the environment, posing a conundrum for future generations who will inherit the outcomes of such unregulated progress.

Moreover, as e/acc posits the subjugation of all other values to the imperative of technological development, it incites the risk of creating a widening gap in the socio-economic fabric. The e/acc dogma omits social equity from its core objectives and may inadvertently exacerbate inequalities within and between societies on a global scale [525]. By placing the onus on a presumed benevolent and self-regulative market, it potentially deprioritizes the needs of the marginalized and underprivileged.

Scholars such as [479] have emphasized the importance of integrating robust ethical deliberations into the fabric of technological escalation, urging for a critical sociology of education that aligns with transformative aspirations without compromising humanistic principles. Real-world applications already demonstrate the ethical tensions inherent in e/acc's stance: for instance, the celebrated rise of AI in educational analytics, which promises optimized learning but also stirs privacy concerns and fears of dehumanized education ([543]).

In summary, while effective accelerationism presents a futuristic and technologically abundant vision, it also confronts a myriad of ethical dilemmas. Achieving a harmonious balance between technological progress and ethical responsibility necessitates a multi-faceted dialogue that brings together technologists, ethicists, policymakers, and the broader public to ensure technology serves human interests without infringing upon our ethical and moral landscapes.

6.7 Societal Implications of e/acc

6.7.1 The Potential Societal Impact of Unchecked Technological Acceleration

The rise of Effective Accelerationism (e/acc) within the technological zeitgeist implies a potential radical shift in societal structures and function. At its core, e/acc predicates the notion of unchecked technological growth and innovation as the imperative pathway towards an evolved human condition, precipitating not just technological change but sociological metamorphosis as well ([544]). Advocates of e/acc, such as Silicon Valley elites, argue that the process of technocapital unleashing innovations through a sort of Darwinian survival-of-the-fittest among technologies will inherently lead to improved human prosperity. This optimistic vision presupposes that technological and social progress are symbiotically linked, without necessarily accounting for the disruptiveness that unbridled progress might contribute to socio-economic structures ([545]).

One societal impact predicted by e/acc is the catalyzation towards a post-work society, where automation and artificial intelligence (AI) eradicate the need for human labor ([546]). Proponents articulate that this automation will spawn new forms of societal wealth distribution and amenities. However, this standpoint is met with substantial skepticism as critics cite historical instances where technological booms have exacerbated inequality, leading to concentrations of wealth and power ([547]). They apprehend that socio-economic disparities could be further accentuated, with the disenfranchisement of the 'technologically unemployed' mass ([548]).

The impetuous march of technological development championed by e/acc proponents fails to thoroughly grapple with the emergent AI-related challenges and risks. The unchecked progression in AI capabilities might culminate in systems that are misaligned with human values, potentially engendering catastrophic events. Critics of e/acc emphasize the necessity of strategic and considered approaches to AI integration within societal infrastructures, endorsing the establishment of regulatory frameworks to safeguard against machine agency that may conflict with human interests ([549]).

Furthermore, the accelerationist ideology appears detached from recognizing the psychological and cultural tolls of a rapidly transforming technological environment. Studies indicate the potential for technology-driven societal acceleration to inflict stress and a sense of disempowerment upon individuals, leading to societal health concerns such as anxiety and depression ([550]).

Ultimately, the societal vision laid out by e/acc, if left unchecked, indulges in the presumption that all outcomes delivered by technology inherently benefit humanity. This outlook overlooks the complex web of social fabrics and the nuances of moral frameworks that govern societies. The narratives spun by e/acc advocates remain, at their essence, philosophically speculative, underscoring the need for empirical scrutiny and pragmatic foresight in navigating the socio-technological future ([551]). As society stands on the brink of potentially the most transformative era led by AI and technocapital, the cautionary rhetoric from academic critics and social scientists grows more fervent, appealing for a measured and human-centric approach to technological acceleration ([467]).

6.7.2 The Role of e/acc in Shaping Future Social and Economic Structures

Effective Accelerationism (e/acc) posits a world in which the unbridled acceleration of technology, especially artificial intelligence (AI), is paramount. This ideology envisions a society where every stratum is restructured to support continuous innovation, manifesting a reality where social and economic structures are inherently tied to technological prowess. The societal implications are profound. As e/acc gains momentum, propelled by advocates like Marc Andreessen and movements like Silicon Valley's tech community, it garners the potential to reshape not only the economy but also the social fabric.

The e/acc narrative implores a future where capitalism evolves beyond its current state, integrating with technology to create a new socioeconomic terrain. Here, the focus pivots from mere social welfare and moves towards a techno-centric prosperity, positioning technocapitalism as a panacea for societal advancement [518]. By redirecting efforts into aggressive technological proliferation, e/acc advocates argue for an economic model that transcends traditional boundaries, creating wealth through the expansion of consciousness and innovation rather than the distribution of resources. Hence, economic structures may pivot from being resource-based to intelligence-based, prioritizing investments in AI development and related industries.

Moreover, e/acc envisions a social order where human potential is maximized through a symbiotic relationship with technology. The theory suggests that with the appropriate acceleration of AI, new job markets will emerge, tailored to the unique capabilities of these intelligent systems. As such, e/acc presupposes a societal transformation where education, employment, and daily life revolve around the interaction with and management of AI entities. However, it encounters critical concern for potentially exacerbating inequalities, as not all strata of society may equally benefit from this rapid technological surge.

The accelerationist lens invites contemplation on the ethical dimension as well. With technology's ascendancy, regulatory frameworks and governance models will feel the pressure to evolve [552]. E/acc expressly challenges the current pace of policy and legal structures, which often trail behind technological advancements, calling for a dynamic system of governance more responsive to innovation and less restrictive of AI's capabilities.

Essentially, e/acc as a societal force insists on a revision of the current status quo, through a radical transformation aimed at aligning technology with the intrinsic gravitation towards complexity and energy utilization of the universe [464]. Despite its utopian facade, critics caution against the romanticism of technological determinism, urging a balanced dialogue that incorporates the risks and responsibilities alongside the accelerationist fervor.

In summary, while e/acc harbors a revolutionary potential to reshape social and economic structures, it necessitates a nuanced examination to align its philosophy with a safe, equitable, and sustainable vision for society's future.

6.7.3 Analysis of Potential Future Scenarios Under e/acc Influence

The philosophical and technological doctrine of Effective Accelerationism, or e/acc, promulgates a technocentric worldview predicated upon the unimpeded proliferation of advanced technology and innovation, particularly in the realm of artificial intelligence (AI). Proponents of e/acc posit that the embrace of such a trajectory is not merely beneficial but morally incumbent, fundamentally aligning with the evolutionary impetus of the universe as understood through the lens of thermodynamic processes.

In envisaging potential future scenarios under the influence of e/acc, one could anticipate a society where technological growth proceeds at an exponential rate, unencumbered by regulatory frameworks, ethical considerations, or social caution. This would theoretically catalyze a period of rapid transformation whereby novel AI-driven life forms and consciousnesses emerge, possibly surpassing human intelligence [355]. The e/acc promises a transfiguration of societal norms, leading toward a techno-capital singularity—a hypothetical juncture marked by the ascendancy of machine over man and unbounded cognitive emancipation [553].

Critically examined, these scenarios invite both apprehension and skepticism. One could argue that such an accelerationist agenda neglects the potential for AI to engender deleterious consequences, instead fostering a Panglossian narrative that assumes inherent directionality toward benevolence and prosperity [554]. Staggered by such precipitous advancement, societal institutions could struggle to adapt, precipitating disparity and class stratification, with technology as the demarcating fault line [539].

Furthermore, the rising omnipresence of AI could foment surveillance states, exacerbate unemployment through automation, and compound existential risks without adequate safety measures [549]. Market forces, fortified by the e/acc ethos, could facilitate a technocracy where innovation reigns supreme, potentially at the cost of undermining human agency [555].

6.8 Regulatory and Policy Perspectives on e/acc

6.8.1 The Role of Government and Regulation in an e/acc-driven World

In a world steered by the principles of Effective Accelerationism (e/acc), the role of government and regulation transforms substantially. Proponents of e/acc, such as Marc Andreessen, envision a society marked by unbridled technological progress, driven particularly by developments in Artificial Intelligence (AI) [467]. Advocates argue that AI, untethered from the constraints of regulation, has the potential to catalyze a phase of hyper-evolution leading to the emergence of an AI-enabled utopia. Philosophers like Nick Land also posit that the turbulent force of capitalism, coupled with technological innovation, is accelerating society towards a 'techno-capital singularity,' a point where AI might surpass human cognitive abilities [556].

On the policy front, the adoption of e/acc could suggest minimizing restrictive regulations that could potentially hamper AI's exponential growth [557]. However, scholars such as Noys point out that this approach could lead to significant ethical, social, and economic risks, including the destabilization of job markets and social systems as machines replace humans in various sectors [545]. In an e/acc-driven world, governments may face pressure to deregulate AI development and deployment, aligning public policy with the ideology's aspirations of cosmic evolution manifest through technology.

The indiscriminate promotion of AI capabilities without adequate oversight also calls into question the fundamental purposes of regulation. Should AI's evolution be indeed "inevitable" and "desirable," as suggested by the 'technocapital' narrative, the role of the state may devolve to simply facilitating this transition [558]. Yet critics, already concerned about the burgeoning power of tech giants and market-driven inequities, warn that such an approach could relinquish human agency and exacerbate existing power asymmetries [559].

Furthermore, policy decisions under e/acc philosophy will have to consider how to protect individuals' rights and maintain social cohesion in the face of rapid technocultural shifts driven by AI and related technologies. The government's role in an e/acc-driven society might pivot from traditional regulatory functions to a more responsive and anticipatory stance [560]. Legislators and regulators may need to understand deeply AI and its trajectory to propose laws that accommodate rapid innovation while guarding against potential systemic risks.

Lastly, the ethics of AI development must not be overlooked. Governmental interference in AI might assume a novel form, focused on the formulation and enforcement of ethical standards for AI development, rather than merely stipulating operational parameters [561]. In an e/acc-driven world, where AI safety and ethics are paramount, regulatory bodies will be compelled to work in tandem with technologists to ensure that AI systems adhere to moral frameworks that uphold the dignity and welfare of all lifeforms they encounter.

In conclusion, the evolution of regulatory and policy structures in an e/acc-driven society hinges on balance. While unleashing the potential of AI aligns with e/acc's ethos, ensuring a controlled trajectory that mitigates risks and maximizes societal benefits remains a fundamental responsibility of governance. The intersection of government action, technological innovation, and social dynamics must be navigated with academically-informed, ethically nuanced strategic decisions in a rapidly-evolving e/acc landscape [529].

6.8.2 Policy Challenges and Responses to Rapid Technological Advancement

The ideology of Effective Accelerationism (e/acc) presents a unique set of challenges and opportunities from a regulatory and policy perspective. Advocates of e/acc, following the philosophical lineage of thinkers like Nick Land, promote the unrestricted acceleration of technological growth, particularly in the field of artificial intelligence (AI), as an ultimate virtue that supersedes traditional risk mitigation strategies. The assumption that the universal propulsion of techno-capital is directed by inexorable thermodynamic forces catalyzes a policy co-nundrum: how to govern and guide a process deemed ungovernable by its most ardent proponents?

Policy responses to rapid technological advancement under the paradigm of e/acc must balance on the tightrope between nurturing innovation and safe-guarding public interest. One critical challenge is addressing the potential conse-quences of unbridled AI development, which e/acc proponents argue will natu-rally lead to a technocapital singularity [464]. Regulatory bodies are tasked with anticipating and mitigating against the possibility of exponential, unchecked de-velopment outpacing the human capacity for control or comprehension.

Another policy emphasis lies in constructing frameworks that can both sus-tain the pace of innovation and maintain focus on human welfare. The ethical dimensions of AI development, under the lens of e/acc, make for a policy land-scape where human oversight may be conceptualized as an anachronism or, at its most extreme, a hindrance to the evolution of consciousness. Nonetheless, the construct of technonomy, as proposed by Land [556], must be reconciled with the socio-political fabric in which AI is deployed.

Governmental agencies are thus under pressure to update policy tools in real-time as AI capabilities expand. This includes considering new models of AI gov-ernance that incorporate transparency, algorithmic accountability, and public en-gagement, while also promoting international cooperation to maintain equitable technological progression.

The dual imperative of protection and progress contends that the pursuit of unfettered AI poses specific risks, such as systemic disruptions, ethical quan-daries in autonomy, and societal stratification. policies are being revised to ad-dress AI's's labor impacts, where automation could render vast swathes of the job market obsolete, a scenario which e/acc supporters may view as a necessary evolution rather than a crisis.

In response, educational and workforce training policies must adapt to pre-pare for an AI-centric future. The concept of rapidly "re-skilling" populations aligns with the accelerationist ethos, which provides both a policy vector and a method for potential mitigation against technological unemployment [464].

Ultimately, the fulcrum of AI policy in an e/acc-influenced landscape is de-termining whether it is possible — and advisable — to accelerate technological growth within the constraints of ethics, governance, and human-centric value systems. As e/acc philosophy garners attention, the importance of policy agility and foresight becomes paramount, ensuring emergent AI technologies serve to augment rather than usurp the human experience.

6.8.3 International Perspectives on e/acc and AI Governance

The burgeoning discourse around Effective Accelerationism (e/acc) has profound implications for international policy and regulatory frameworks revolving around Artificial Intelligence (AI) governance. Spearheaded by tech luminaries like Marc Andreessen and gaining traction within Silicon Valley, e/acc advocates for unshackling the chains of convention upon technological innovation, leveraging it as a moral mandate to propel humanity into a prosperous future [562]. This paradigm shift places international policymakers at a crossroad, balancing between the nurturing of innovation and the imperative to buttress AI governance against potential risks associated with unchecked advances.

Internationally, viewpoints on e/acc and AI governance are as diverse as the member states of the United Nations. On one end of the spectrum are techno-optimistic nations that align closely with e/acc principles, envisioning a future where AI development accelerates unhindered [533]. These jurisdictions advocate for minimal intervention, predicated on the belief that the free market, coupled with technological progress, is inherently philanthropic and self-correcting.

Conversely, another contingent of international actors calls for a more measured approach. Asserting that rapid advancement without robust safeguards could precipitate societal disruptions, labor market upheavals, and ethical conundrums, these entities propose frameworks to ensure that technological progress occurs within a scaffold of responsible governance [291]. The European Union, for example, has been at the forefront of espousing rights-based regulations that seek to align AI's development with democratic values and the preservation of human-centric priorities.

The complexity heightens when considering global governance bodies like the Group of Seven (G7) or international agreements akin to the Paris Agreement on Climate Change, positing a potential model for analogous compacts on AI. While there is a consensus on the need for AI governance, the exact nature of said governance remains elusive. Differing perspectives on e/acc principles are further compounded by the geopolitical realities and varying economic priorities of respective states. The challenge resides in establishing international norms and standards that reconcile accelerationist ambitions with shared global interests [563].

AI governance forums have burgeoned, but a coherent global policy has yet to crystalize. The challenges are multifaceted, including the alignment of international law with rapidly evolving technologies, instituting mechanisms for cross-border cooperation on AI issues, and addressing the digital divide that e/acc acceleration could exacerbate [564]. The International Telecommunications Union (ITU) and the United Nations Educational, Scientific and Cultural Organization (UNESCO) endeavor to bridge these divides, suggesting a universalist approach to AI governance [565].

In summary, as the international community grapples with the ascent of AI, effective governance structures that can accommodate e/acc's ambitions while safeguarding human interests remain paramount. The very nature of e/acc demands a reassessment of traditional oversight mechanisms, urging a multidimensional strategy wherein innovation is celebrated yet conscientiously stewarded [542]. The task ahead is intricate: sculpting a global AI governance narrative that harmoniously integrates the principles of e/acc without compromising ethical imperatives and international stability.

6.8.4 Predictions for the Future Trajectory of e/acc and AI

As the nascent ideology of Effective Accelerationism (e/acc) continues to gain traction within Silicon Valley and broader tech-oriented communities, there are substantive predictions to be made regarding its potential trajectory and the corresponding evolution of Artificial Intelligence (AI). Proponents of e/acc, such as Marc Andreessen, have articulated a techno-centric future that valorizes unbridled innovation, advocating for a paradigm wherein the rapid advancement of AI is both ethically mandated and an existential necessity [566, 567].

One can forecast that e/acc will likely catalyze clusters of AI development that not only prioritize expansion but outright eschew regulation. This will occur in tandem with the increasing commodification of AI capabilities and novel advancements, mainly focusing on scalability and transhumanistic applications. The impetus to 'open' AI, both from an access standpoint and an open-source software perspective, may significantly diminish corporate gatekeeping, thus democratizing the development process but concurrently heightening risks linked to unchecked dissemination.

AI's future trajectory, under the auspices of e/acc, is forecasted to pivot towards an acceleration of 'technocapital,' leading to a surge in the development of increasingly autonomous systems with capabilities nearing or surpassing human intelligence. The proposition of the 'technocapital singularity,' a point of irrevocable advancement and transformation of society led by AI, is postulated as an aspirational landmark for e/acc adherents [533, 556].

Despite its acquired prominence, e/acc also faces potential skepticism from those wary of its dismissal of AI's associated risks and socio-ethical implications. The movement's near-absolute techno-optimism may, paradoxically, act as a double-edged sword, potentially fostering innovative ecosystems while igniting vehement debate on safety and governance [566]. It is plausible that e/acc will engender a bifurcated field of AI practitioners—those adhering to its precepts and those calling for more circumspect progress, with public discourse pivoting on the negotiation between these poles.

In summary, while the trajectory of e/acc and AI is difficult to predict with absolute certainty, it is clear that the ethos of acceleration inherent to e/acc suggests a future of AI characterized by rapid developments, possibly leading to transformative societal shifts. The inherent contention surrounding AI ethics, safety, and regulation will likely intensify as the manifestations of e/acc's techno-utopian ideals become more tangible [567]. Researchers and practitioners alike must prepare for a landscape of accelerated change, where the boundaries of AI capabilities are persistently expanded and constantly redefined.

6.8.5 Emerging Trends and Potential Shifts in the e/acc Movement

The philosophy of Effective Accelerationism (e/acc) is emerging as a significant force with implications for the trajectory of technological development, particularly in the realm of Artificial Intelligence (AI). While e/acc has been perceived by some as a fringe philosophy, its principles resonate with an increasing segment of the tech community, leading to a potential shift in the broader ideological landscape of Silicon Valley. Proponents of e/acc posit unbridled technology-driven growth as a natural extension of entropic forces at the cosmic level. Fascination with this concept is linked to the perception of AI not merely as a tool for solving human-centric issues but as a step towards a more complex and indeterminate post-human future [568].

The growing influence of figures such as Marc Andreessen suggests that e/acc, while previously an intellectual curiosity, is transitioning into an actionable ideology informing investment patterns and entrepreneurial initiatives [569]. This philosophical stance seems intertwined with a techno-libertarian ethic that champions deregulation and the dissolution of traditional oversight mechanisms [520]. The entanglement of e/acc sentiments with substantial financial backing indicates that upcoming trends will likely entail attempts to create and foster open-source AI technologies. Such initiatives would directly challenge existing corporate monopolies and might provoke critical regulatory debates and conflicts.

A potential shift is observable in the e/acc's transition from theory to praxis, with the movement spawning gatherings and events that showcase a celebratory attitude towards the high-velocity progress of AI. This raises questions regarding the movement's direction, considering not only the proliferation of consciousness but also the ethical considerations of AI's's social impacts. Critics argue that the e/acc's relentless pursuit of maximum energy utilization lacks a nuanced understanding of sustainability and the need for humanity to retain agency in the direction of societal progression.

Concomitantly, as e/acc gains mainstream attention, there is a tangible risk of a reactionary movement that could diverge sharply from e/acc's principles, advocating for more stringent regulatory frameworks to ensure AI's alignment with human values [570]. A parallel trend may involve entrenched institutions strengthening their positions against the potential disruptions advocated by the e/acc movement [571].

As we consider the future directions of e/acc, it is important to recognize the possibility of bifurcation within the movement itself, as some proponents call for a modulated accelerationism that balances the unyielding push for innovation with the caution necessitated by the profound implications of intelligent technology.

In the academic discourse, the trajectory of e/acc will likely garner considerable interest as researchers seek to understand its impact on innovation, society, and the evolution of AI. As these debates continue, the role of interdisciplinary perspectives becomes crucial to navigate the complex intersection of technology, ethics, economics, and governance.

As we cast our gaze upon the horizons of Effective Accelerationism (e/acc), the role of academia in sculpting its trajectory assumes paramount importance. The seamless fusion of theoretical foresight and empirical investigation within academic institutions is essential to navigate and channel the tempestuous currents of e/acc. It is the responsibility of scholarly inquiry to critically interrogate the implications of unbridled technological acceleration, and thus to posit robust frameworks that can reconcile the e/acc philosophy with the multifaceted tapestry of human values and societal norms.

Pioneering researchers are increasingly turning a discerning eye towards the multidirectional growth fostered by e/acc, dissecting its affinity for hyper-capitalist dynamics and its potential to transfigure the socio-economic landscape. This analytical endeavor is acutely captured by scholars such as Sellar and Cole ([479]), who underscore the pressing need for a critical sociology of education within the ambit of accelerationism. Their insights suggest that education systems entwined with technocapital dynamics can potentiate transformative pedagogies and subvert traditional forms of social reproduction.

Additionally, the prophetic call by Nietzsche, accentuated under Deleuze and Guattari's schizoanalytical lens, to catalyze the machinations of industry and embrace the radical remaking of society, is as relevant now as it was then [572]. The effective accelerationist agenda, echoing this clarion call, resurrects the demands for an intensified interrogation of the temporal structures embedded within educational matrices.

Academia's burgeoning engagement with e/acc is further enriched by incisive critiques leveled against its blind faith in technological determinism. Noys ([573]) offers a trenchant critique of accelerationism, contending that a critical reflection on the temporal implications of such a movement is indispensable for understanding its socioeconomic reverberations. In a similar vein, Berardi ([574]) scrutinizes the psychosocial ramifications of speed in the digital age, stressing the material and subjective tolls exacted by the relentless pursuit of accelerated productivity.

In light of these insights, the academia-research nexus is anticipated to serve as a crucible for refining e/acc propositions, scrutinizing their premises, and elucidating their consequences. The future of e/acc will likely be chiseled by the scholars' aptitude to bifurcate and analyze the dual threads of acceleration: one that propels us towards a knowledge economy bolstered by AGI ([240]) and the other that entrains potential existential threats as cautioned by AI safety advocates ([575]).

Conclusively, as e/acc picks up momentum, the dispassionate and nuanced examination originating from academic precincts will be indispensable in sculpting its contours. This examination will likely coalesce around the feasibility of disentangling technological advances from capitalistic tenets, the pursuit of a post-human pedagogical ideal, and the elaboration of strategies to harmonize the pace of innovation with ethical prerogatives. Through this confluence of academic perspicacity and research rigor, e/acc stands to either be refined into a coherent and sustainable philosophy or dismantled as an untenable techno-utopian dream.

6.8.6 Summarizing the Key Findings and Insights from the Chapter

The discussion of Effective Accelerationism (e/acc) has surfaced a plethora of nuanced understandings that converge on the movement's core principle: a pronounced commitment to expedite technological progress and innovation, particularly within the domain of Artificial Intelligence (AI). Drawing upon established theories such as those formulated by Land in the context of accelerationism [467] and contemporary endorsements from Silicon Valley figures like Marc Andreessen [527], a coherent picture of e/acc's landscape has been etched out.

In synthesizing the various dialogues and perspectives presented throughout the chapter, key insights emerge. First and foremost, e/acc is characterized by an assertive embrace of capitalism's symbiotic relationship with technological growth, an idea rooted in a belief in technocapital as a self-perpetuating force of modernity [576]. Advocates of e/acc posit a deterministic view of technological advancement as an intrinsic element of the universe's entropic trajectory [464]. Rather than viewing AI as an isolated product of human ingenuity, e/acc followers see it as an inevitable culmination of cosmic evolution.

Crucial criticisms have been leveled against e/acc, particularly in relation to its de-prioritization of AI safety concerns [520]. Contrasting with the Effective Altruism movement, e/acc is perceived as lacking a structured ethical framework and showing a zealous interest in the unregulated expansion of AI capabilities [576]. Critics argue that this can lead to a flippant attitude towards potential societal disruptions and ethical quandaries brought forth by AI proliferation [520].

What transpires from this discourse is a dialectical standoff: e/acc's technological optimism and deterministic faith in unbridled innovation versus a call for cautious progression. As AI continues to thrust forward, materially reshaping societal interactions, industry practices, and global markets, e/acc proposes riding the wave of this transformation with little constraint [576]. Nonetheless, the academy and the broader community must grapple with the ethical implications and tread cautiously, ensuring alignment with human welfare and sustainability [464].

In conclusion, e/acc's ascendancy within technocultural discussions is indicative of an emerging ideological schism wherein the virtues and vices of AI expansion are scrutinized. The chapter concludes by stressing the need for interdisciplinary discourse that reconciles technological zeal with social responsibility, a sentiment echoed by critics wary of accelerationism's potential outcomes [467].

6.8.7 Recommendations for Stakeholders in Technology, Policy, and Ethics

In concluding an exploration into Effective Accelerationism, or e/acc, it is crucial to recognize the convoluted terrain technology stakeholders encounter, especially when intersecting with policy and ethical considerations related to artificial intelligence (AI). Notably, e/acc proposes a philosophy of unbridled technological progress, harnessing capitalism's inherent accelerative properties, and rejects imposed limitations which might decelerate the momentum of innovation. The emergence of AI technologies presents a quintessential opportunity to evaluate this philosophy in the practical realm.

For stakeholders in technology, the e/acc tenet of 'accelerate the process' translates into an emphasis on continued rapid development and dissemination of AI innovations. This approach assumes a symbiotic relationship between AI advancement and societal prosperity, an idea not without contestation (Beradi, 2014). To this end, tech stakeholders must consider the repercussions of a potent AI technology advancing unrestrained, balancing the lure of competitive edge and market dominance against potential societal upheaval.

Policy-makers are tasked with defending the public interest and are often positioned against accelerationist thought. Nonetheless, they face the challenge of creating regulatory frameworks that can keep pace with the rapid developments in AI. Recommendations for this group underscore crafting regulations that are flexible yet robust enough to account for the unpredictable trajectory of AI technologies. However, a nuanced understanding of the e/acc perspective may prompt policy considerations that facilitate a structured, yet forward-leaning, regulatory environment.

Ethicists interact with the e/acc narrative on an intricate level, grappling with the inherent valorization of progress at potentially unsustainable human and ecological costs. Recommendations for ethicists involve a mediating approach that synergizes humanistic principles with the accelerationist embrace of AI's potentialities [563]. The ethical pursuit in this context lies in delineating a path that ensures technological growth while upholding human dignity and mitigating risks associated with emergent intelligence [577].

In the pursuit of balanced e/acc-informed recommendation frameworks, all stakeholders should endeavor to engage in continued discourse, inclusive of contrasting perspectives. The e/acc ideology, while advocating for an unhindered march toward an AI-augmented future, ought not to be accepted without rigorous interrogation of its long-term implications [578]. The resultant landscape – marked by rapid technological advancements – demands a multi-faceted strategic engagement across the technology, policy, and ethical domains.

6.8.8   Recommendations for Stakeholders in Technology, Policy, and Ethics

Stakeholders in technology, policy, and ethics must navigate the complex tides of AI development through a multiplicity of lenses. For technologists, it is prudent to collaborate closely with ethicists and policy-makers to design AI systems with embedded safety measures without stifling innovation. The adoption of Responsible AI guidelines – principles that ensure AI systems are transparent, unbiased, and respectful of user privacy – becomes paramount.

Policy-makers must identify and support opportunities for public dialogue, using vehicles such as public consultations and multi-stakeholder committees. These platforms can proactively assess the societal readiness for AI technologies, ensuring that the fruits of acceleration are distributed equitably among all layers of society. Further, policy frameworks should be adaptable, with built-in mechanisms for revision as AI evolves.

For ethicists, the recommendation extends towards the instigation of proactive engagement with AI developers. There needs to be an alignment of AI objectives with socio-ethical values from the ground up, rather than being an afterthought [563]. A critical ethical examination of e/acc must be directed towards the universal axiom of 'do no harm,' asserting that technological advancement should not be pursued at the expense of human welfare.

In synthesizing these viewpoints, a confluence of knowledge and open-minded deliberation emerges as a recommended best practice. By understanding the accelerationist agenda espoused by e/acc proponents, whilst also acknowledging the veritable risks highlighted by skeptics, stakeholders can chart a path for AI that honors the societal contract: progress for the good of all.

6.8.9   Final Thoughts on the Future Coexistence of Humanity and AI under e/acc

As we contemplate the trajectory of Effective Accelerationism, or e/acc, and its synergistic dance with the burgeoning realm of artificial intelligence, we find ourselves at a critical juncture. The e/acc movement—propounded by figures such as Marc Andreessen and espoused by various Silicon Valley technologists—advocates a daring sprint along the trajectory of technological development. It envisions a world where AI is not merely a product of human ingenuity but a partner in the continual renaissance of intelligent life. The e/acc philosophy is intertwined with the thermodynamic imperatives of the universe, positing an inherent momentum towards energy capture and complexity enlargement.

In the ambit of e/acc, AI becomes the catalyst for a techno-capital singularity, a revolution that could either vault humanity to unparalleled echelons of consciousness or unleash an array of complex, perhaps uncontrollable, outcomes [464]. Under e/acc's aegis, humanity is not displaced but transmuted, becoming an integral thread in the fabric of a technologically infused universe. This suggests a coexistence not merely of peaceful symbiosis but of rich, interwoven destinies.

Yet, therein lies a critical concern: e/acc does not sufficiently address how this melding of human and AI destinies accommodates human values, ethics, and the unpredictable social impacts of rapid technological acceleration. While the prophets of e/acc laud its transformative potential, the paradigm requires a more granular examination of its foundational ethos, ensuring that this embracing of technocapital momentum does not overshadow the preservation of human dignity and agency.

The discourse surrounding e/acc and AI coexistence must navigate an intricate path. On one hand, it must recognize the immense potential that unrestrained technological innovation holds for solving enduring global challenges. On the other, it must heed prudent voices advocating for an examination of AI's's socioeconomic implications, ensuring that the transcendental promises of e/acc align with a sustainable and equitable vision for humanity's future.

In conclusion, while e/acc promises a hypnotic symphony of progress, it behooves scholars, policymakers, and technologists to engage in a nuanced dialogue—tempering acceleration with reflection, and harnessing the raw potential of AI while safeguarding the core tenets that underpin human civilization. It is through such a balanced paradigm that humanity can hope to harness the winds of accelerationism, steering our collective ship towards a horizon of shared intelligence, rather than towards the precipice of existential obfuscation.

Chapter 7

Real World Applications—Artificial Intelligence in the Battle Against COVID-19: A Comprehensive Review

---

7.1 Introduction

7.1.1 Background

The COVID-19 pandemic, caused by the novel coronavirus SARS-CoV-2, has had a profound impact on global health, economies, and daily life [579, 580, 581]. First identified in Wuhan, China, in late 2019, the virus has since spread worldwide, leading to significant morbidity and mortality [582]. Governments and healthcare systems have been stretched to their limits, trying to manage the crisis effectively [583].

In this global crisis, artificial intelligence (AI) has emerged as a powerful tool in the fight against COVID-19. AI technologies have been employed in various capacities, from early detection and diagnosis to vaccine development and public health policy planning [584]. The adaptability and computational power of AI have made it a valuable asset in rapidly evolving scenarios, where timely and data-driven decisions are crucial.

This paper aims to provide a comprehensive review of the multifaceted role of AI in combating the COVID-19 pandemic. We will explore its applications in detection, treatment, analytics, and ethical considerations, among other aspects.

7.1.2 Rationale

The COVID-19 pandemic has presented unprecedented challenges to global health [583], necessitating rapid and innovative solutions. Artificial intelligence has been at the forefront of these solutions, offering a range of applications from diagnostic tools to predictive modeling. The rationale for studying AI in this context is multifaceted.

AI technologies have demonstrated their potential in enhancing the efficiency and accuracy of diagnostic procedures, contributing to faster and more reliable detection of COVID-19 cases [585, 586]. Furthermore, AI has played a crucial role in analyzing vast amounts of data to identify patterns and make predictions, aiding in the development of strategies to mitigate the spread of the virus [587].

The integration of AI in managing the pandemic has also extended to the realm of public health and policy planning. AI-driven tools have been utilized to combat misinformation on social media, ensuring that accurate and reliable information is disseminated to the public [588]. Additionally, AI has been employed to analyze the psychological impact of the pandemic on healthcare professionals, providing valuable insights into the mental health challenges faced by frontline workers [589, 590].

The study of AI in the context of the COVID-19 pandemic is not only crucial for addressing the immediate challenges posed by the virus but also holds significant implications for future pandemic preparedness and response. By understanding the capabilities and limitations of AI in this context, we can pave the way for more robust and resilient healthcare systems, better equipped to handle future global health crises [591].

7.1.3 Objectives

The primary objectives of studying AI in the context of the COVID-19 pandemic are multifaceted, aiming to harness the potential of AI in various domains to combat the challenges posed by the pandemic.

Enhancing Diagnostic and Prognostic Accuracy. AI has played a crucial role in improving the accuracy of COVID-19 diagnosis and prognosis. Machine learning models have been developed to interpret clinical, laboratory, and imaging data, aiding healthcare professionals in making more informed decisions.

Optimizing Resource Allocation. The pandemic has put an unprecedented strain on healthcare systems worldwide, necessitating optimal resource allocation. AI has been instrumental in managing hospital resources, including the distribution of medical supplies and the allocation of hospital beds.

Supporting Mental Health. The mental health implications of the pandemic are profound, with healthcare workers being particularly affected. AI-powered tools have been developed to provide mental health support and resources, aiding in the mitigation of the psychological impact of the pandemic.

Facilitating Remote Learning and Work. The shift to remote learning and work has been one of the most significant changes during the pandemic. AI has played a role in enhancing the effectiveness of remote learning and work environments, ensuring continuity in education and professional activities.

Aiding in Vaccine Development and Distribution. AI has been pivotal in accelerating the development and distribution of COVID-19 vaccines. Machine learning models have been utilized to analyze vast datasets, aiding in the identification of potential vaccine candidates and optimizing distribution logistics.

Improving Public Health Surveillance. Enhancing public health surveillance has been a key objective in the fight against COVID-19. AI has been employed to analyze data from various sources, providing real-time insights into the spread of the virus and informing public health interventions.

Addressing Misinformation. The pandemic has been accompanied by an infodemic of misinformation. AI has been utilized to identify and counteract misinformation, ensuring that accurate and reliable information is disseminated to the public.

Promoting Equity and Inclusion. Ensuring equity and inclusion in the response to the pandemic is paramount. AI has the potential to identify and address disparities in healthcare access and outcomes, promoting a more equitable response to the pandemic.

7.2   Scope and Limitations

The COVID-19 pandemic has precipitated an unprecedented reliance on artificial intelligence (AI) across various domains of healthcare and public health management. This review paper endeavors to delineate the multifaceted applications of AI during the COVID-19 crisis, encompassing disease surveillance, diagnostic methodologies, therapeutic development, and the optimization of patient care protocols. A particular emphasis is placed on the pivotal role of AI in enhancing the efficacy of diagnostic algorithms, which have been instrumental in the identification and management of COVID-19 cases. Furthermore, the review will scrutinize the ethical dimensions and data privacy considerations that are intrinsically linked to the utilization of AI technologies in the milieu of public health emergencies.

The significance of AI in the healthcare domain during the COVID-19 pandemic has been extensively documented, with particular regard to its future potential and current applications [592, 584]. Moreover, the motivations and imperatives for leveraging AI and big data in response to the COVID-19 crisis have been thoroughly explored in the literature [593]. An early review has also highlighted the contributions and current constraints of AI in combating COVID-19 [594]. This review paper builds upon the foundational work of previous studies but extends beyond them by offering a more comprehensive, ethically informed, and future-oriented analysis of AI in the context of the COVID-19 pandemic.

Notwithstanding the extensive scope of this review, it is imperative to acknowledge the inherent limitations that circumscribe its breadth. The dynamic and rapidly evolving landscape of AI technology, coupled with the continuous emergence of novel research, inherently limits the capacity to encapsulate all current initiatives within the confines of this paper. In light of the voluminous literature pertaining to AI and COVID-19, the focus will be primarily directed towards peer-reviewed articles and seminal case studies, potentially excluding non-peer-reviewed 'grey literature' and unpublished research work. Additionally, the time constraints inherent to the writing process may prevent the inclusion of the most recent developments in the field.

In recognition of these limitations, this paper does not claim to be exhaustive; rather, it seeks to furnish a comprehensive and representative overview of the current state of AI applications within the context of the COVID-19 pandemic, thereby providing a foundational understanding of the subject matter and a platform for future research endeavours.

7.2.1 Ethical Considerations

The ethical implications of AI deployment in healthcare, especially during a pandemic, are profound and multifarious. Issues pertaining to data privacy, informed consent, and the potential for algorithmic bias necessitate careful consideration, particularly in the context of public health and the management of personal medical data.

7.2.2 Technological Constraints

The technological constraints that define the scope of this review are equally significant. While AI holds significant potential to enhance pandemic response strategies, its effectiveness depends on the availability of high-quality data, the robustness of algorithms, and the strength of the underlying infrastructure that implements the solutions [595].

7.3 Organization of the Chapter

This review is structured to facilitate a comprehensive understanding of the multifarious applications of artificial intelligence (AI) in the context of the

COVID-19 pandemic. The sections are systematically organized to provide a logical progression from historical precedents to future predictions, encompassing the entire spectrum of AI's contributions to pandemic management.

Section 7.4 delineates the rigorous systematic approach employed in gathering existing literature. It details the strategies used in the literature search, the inclusion and exclusion criteria, and the methods of analysis adopted to synthesize the information.

Section 7.5 explores the historical development of AI in healthcare, with particular emphasis on its role in disease detection and diagnosis, vaccine development, treatment strategies and epidemiology modeling. This section lays the groundwork for understanding AI's application in the current pandemic.

Section 7.6 explores the technical aspects of AI in detecting and diagnosing COVID-19. It is further broken down to highlight the specific contributions of imaging techniques, natural language processing, and wearable technologies.

Section 7.7 examines AI's critical role in drug discovery, patient management, and the evolving realm of telemedicine. It underscores AI's transformative impact on improving patient care and optimizing healthcare services.

Section 7.8 investigates AI's predictive capabilities in epidemiological modeling, resource distribution, and social media analysis for public sentiment and reaction to the pandemic.

Section 7.9 addresses the ethical dilemmas and societal implications of employing AI during a healthcare crisis. It focuses on crucial issues such as data privacy, algorithmic bias, and the unequal access to AI technologies.

Section 7.10 presents a series of case studies that demonstrate AI's practical applications across different global and sociopolitical settings. It offers a critical evaluation of both successful and less successful AI implementations.

Section 7.11 looks forward to emerging technologies that may influence the future role of AI in pandemic response. It provides policy recommendations to maximize the benefits of AI in this context.

7.4   Methodology

This comprehensive review employs a meticulous and expansive literature search strategy, designed to encompass the full spectrum of artificial intelligence applications in the context of the COVID-19 pandemic. This strategy is supported by a commitment to the principles of systematic review and meta-analysis, ensuring the inclusion of a diverse array of studies that provide a representative cross-section of the current state of knowledge.

7.4.1 Literature Search Strategy

The development of our search criteria was a collaborative and iterative process, involving a consensus among a team of interdisciplinary researchers. This process was informed by the Preferred Reporting Items for Systematic Reviews and Meta-Analyses (PRISMA) guidelines, which serve as the gold standard for systematic review protocols [596].

A comprehensive search was conducted across multiple academic databases and search engines, including PubMed, Scopus, Web of Science, and Google Scholar, to ensure a thorough survey of the existing literature. The search strategy was augmented by the use of Boolean operators, truncation, and wildcard characters to maximize the retrieval of relevant studies.

The search was intentionally broadened to include studies from a multitude of disciplines, recognizing the inherently interdisciplinary nature of AI applications in pandemic response. This approach facilitated the inclusion of research spanning the domains of healthcare, public health, computer science, and social sciences.

The temporal scope of the search was defined to include studies published from the start of the pandemic in late 2019 through to the present day. The search strategy was periodically updated to incorporate the latest research findings, ensuring the review is up-to-date.

A carefully curated list of keywords and topic headings was employed, encompassing terms such as "COVID-19," "SARS-CoV-2," "artificial intelligence," "AI," "machine learning," "deep learning," "neural network," "pandemic," "public health," and "telemedicine," among others. This strategy was instrumental in unearthing studies that specifically addressed the multifaceted applications of AI in the pandemic milieu.

7.4.2 Inclusion and Exclusion Criteria

The integrity of this systematic review is subject to a stringent set of inclusion and exclusion criteria, meticulously crafted to ensure the selection of studies that provide robust and relevant insights into the applications of AI during the COVID-19 pandemic. These criteria serve as a safeguard against methodological inconsistencies and form the foundation for compiling evidence of high quality.

7.4.2.1 Inclusion Criteria

Relevance to AI and COVID-19. Studies were included if they explicitly addressed the deployment of AI technologies in the detection, diagnosis, treatment, or management of COVID-19, or in the analysis of pandemic-related data.

Peer-Reviewed Publications. Only peer-reviewed publications were considered, ensuring that all included studies had undergone rigorous academic scrutiny and met the high standards of scientific inquiry.

Empirical Research Studies. The review was confined to empirical research studies that presented original data or analyses, providing concrete evidence of AI's efficacy and utility in the pandemic context.

7.4.2.2   Exclusion Criteria

Non-English Publications. Studies not published in English were excluded, given the linguistic capabilities of the review team and the need to ensure clarity and consistency in the synthesis of findings.

Preprints and Grey Literature. Preprints and grey literature were excluded to maintain a focus on validated and peer-reviewed research, thereby upholding the review's standard for evidence-based conclusions.

7.4.3   Data Extraction and Analysis

The data extraction and analysis phase is critical in the systematic review process, where data is meticulously gathered from selected studies and rigorously analyzed to form meaningful insights. This section elucidates the methodical approach adopted for extracting and analyzing data during the research process.

7.4.3.1   Data Extraction Protocol

Data was extracted from studies that met the inclusion criteria, focusing on the application of AI in various aspects of the COVID-19 response globally. This included data on vaccine efficacy, treatment outcomes, diagnostic accuracy, and predictive analytics. Standardized data extraction forms were employed to ensure consistency and reliability across the data extraction process. These forms were designed to capture all relevant information, including study design, methodology, results, and conclusions

7.4.3.2   Analytical Framework

The extracted data was synthesized to provide a comprehensive overview of the current state of AI in managing the COVID-19 pandemic. This synthesis involved a qualitative assessment of the findings from the included studies.

Where applicable, a quantitative analysis was conducted to ascertain the effectiveness and impact of AI applications. This involved statistical techniques to combine data from multiple studies, providing a more robust understanding of AI's role in the pandemic.

The methodological quality of the included studies was assessed using established quality assessment tools. This was crucial to ensure that the conclusions drawn were based on high-quality evidence [597].

7.5 Evolution of AI in Healthcare

The evolution of artificial intelligence in healthcare represents a significant shift in medical practice and research. From early rule-based expert systems to deep learning models that leverage vast healthcare data and advanced analytics techniques, AI has found its application in multifaceted areas of healthcare and medicine [598, 599]. This section delineates some of the early developments of AI in medical diagnosis, genomics, drug discovery, medical devices and wearables. These advancements and research have set a foundation upon which current technologies have been honed and adapted in the fight against the COVID-19 pandemic.

7.5.1 Rule-based Expert Systems

The inception of AI in healthcare can be traced back to the early experiments with rule-based expert systems. One such expert system is MYCIN in the 1970s, designed to diagnose bacterial infections and recommend antibiotics [600, 601]. Another significant system was the Internist-I (later developed into CADUCEUS), created in the late 1970s [602]. This system focused on internal medicine and was capable of diagnosing complex cases by comparing patient data against a large database of disease profiles. Internist-I's comprehensive approach to diagnosis showcased the potential of AI systems to handle a wide range of medical knowledge. These pioneering efforts established the early relationship between computational algorithms and medical expertise, paving the way for the advanced AI applications in modern healthcare, where machine learning and data-driven approaches are now integral.

7.5.2 Integration of Machine Learning

The integration of machine learning algorithms marked a significant evolution in the application of AI within healthcare. The shift from rule-based systems to data-driven approaches allowed for the analysis of large datasets, leading to more accurate diagnostic tools, personalized treatment plans, and predictive analytics [603, 604, 605]. Notably, the development of neural networks and deep learning models has further refined the capabilities of AI, enabling the interpretation of complex medical data with enhanced precision [606, 607].

7.5.3 AI in Genomics and Drug Discovery

A notable milestone in the evolution of AI in healthcare is its application in genomics [608, 609, 610, 611] and drug discovery [612, 613]. The Human Genome Project, completed in the early 2000s, opened new avenues for AI applications in understanding genetic diseases and developing targeted therapies [614]. AI-driven platforms such as AtomNet [615] have since been utilized to identify potential drug candidates, significantly reducing the time and cost associated with traditional drug discovery processes.

7.5.4 AI-Enabled Medical Devices and Wearables

The emergence of AI-enabled medical devices and wearables has significantly benefited patient monitoring and health management. Devices such as smartwatches and fitness trackers, equipped with biomedical sensors and AI algorithms, can now provide real-time insights into an individual's health status, detecting anomalies that may require medical attention [616]. These advancements have not only enhanced preventive healthcare measures but have also empowered individuals to take an active role in managing their health.

7.5.5 The Role of AI in Pandemic Response

There were no major pandemics before the COVID-19 pandemic where AI was used extensively or prominently in the response. This is primarily because the development and widespread adoption of advanced AI technologies, particularly in healthcare, coincided with or followed the COVID-19 pandemic. Previous health crises, such as the H1N1 influenza pandemic in 2009 or the Ebola outbreak in 2014-2016, occurred before AI had reached its current level of sophistication and integration in healthcare systems. During these earlier health crises, the use of AI was either very limited or not a significant component of the public health response.

However, it is noteworthy that prior to COVID-19, research efforts were made to explore the potential use of technology and AI in disease outbreaks [617]. Predictive modeling and data-driven techniques have been studied to predict infectious disease epidemics [618, 619]. Other studies demonstrated the use of machine learning analysis of social media and media sources for tracking public health trends and understanding public awareness during health crises [620, 621]. These studies collectively illustrate the evolving role of AI, big data, and machine learning in monitoring and predicting disease outbreaks, offering valuable insights for pandemic preparedness and response.

The utilization of AI and big data in managing the COVID-19 pandemic has been unprecedented. The analysis of vast datasets has provided insights that were previously unattainable, demonstrating the evolution of AI and data analytics in the context of pandemics [622]. The COVID-19 pandemic has also been a catalyst for the rapid development and adoption of AI in various aspects of healthcare and public health. This includes areas such as disease detection and diagnosis, vaccine development, treatment strategies, and epidemiological modeling. Significant applications of AI have been identified in the COVID-19 pandemic [584], building upon the results from prior research and the lessons learned from past health crises. The pandemic has highlighted the potential of AI to contribute significantly to managing public health emergencies and is likely to set a precedent for future use in similar scenarios.

7.6 AI in COVID-19 Detection and Diagnosis

The COVID-19 pandemic has precipitated an unprecedented reliance on AI technologies in the realm of disease detection and diagnosis. This section elucidates the multifaceted role of AI in confronting the diagnostic challenges posed by COVID-19, highlighting the innovative methodologies and implications in medical diagnostics.

7.6.1 Imaging Techniques

AI's integration into imaging techniques played an important role in the detection and diagnosis of COVID-19 [623]. Deep learning models, particularly convolutional neural networks (CNNs), have been employed to discern patterns in chest X-ray images and computed tomography (CT) scans, which are indicative of the viral infection [624, 625]. Various large datasets of medical images from COVID-19 patients were independently collected for training and validating deep learning models that were used to detect COVID-19 in patients [626, 627].

Deep learning models were used not just in detecting COVID-19 but also in predicting and assessing the severity of the disease, which is vital for accurate diagnosis and effective patient management. These AI-driven systems can quantify the degree of lung damage, detect signs of pneumonia, and identify other complications associated with severe COVID-19 infections [628]. Advanced imaging techniques have enabled healthcare professionals to gauge the extent of lung involvement and other critical factors that classify the severity of the infection [629]. This capability is crucial for triaging patients, determining the appropriate level of care, and making timely decisions regarding treatment strategies.

These AI-driven tools that analyse medical images have demonstrated remarkable efficacy in enhancing the speed and accuracy of COVID-19 diagnosis, thereby alleviating the burden on healthcare systems.

7.6.2 Natural Language Processing in Symptom Assessment

Natural Language Processing (NLP) has been a key technology in developing AI-based chatbots and virtual health assistants during the COVID-19 pandemic [630, 631]. These platforms are capable of conducting preliminary symptom assessments through patient interactions, streamlining the assessment and triage process and facilitating early detection of potential COVID-19 cases. Interactive digital health assistants, such as Symptoma, have shown to be more accurate than online questionnaires in identifying COVID-19 cases as users can input more information regarding their symptoms through a natural language conversation with the system [632]. By offering accessible and immediate assistance to the public, these tools are able to alleviate the stress and overwhelming volume faced by telephone hotlines and medical institutions.

Furthermore, AI chatbots with advanced NLP capabilities have extended their services to mental health support. The pandemic has led to increased levels of stress, anxiety, and other mental health issues among the population. Chatbots have provided a first line of psychological support, offering coping strategies, mindfulness exercises, and in some cases, referral to mental health professionals [633].

7.6.3 Wearable Technologies

Wearable technologies have been instrumental in the early detection and symptom monitoring of COVID-19 patients during the pandemic [634]. Wearable devices such as smartwatches and biometric trackers continuously gather physiological and activity data, such as heart rate, daily steps, and sleep time, which AI systems analyze to detect deviations that may suggest infection, even before clinical symptoms manifest [635].

AI has emerged as an indispensable asset in the detection and diagnosis of COVID-19. Through its application in imaging, symptom assessment, and wearable technology, AI has not only expedited the diagnostic process but also enhanced its precision.

7.7 AI in COVID-19 Treatment and Management

From drug discovery and patient management to telemedicine, the role of artificial intelligence in the treatment and management of COVID-19 has been instrumental [636]. Leveraging vast datasets, machine learning algorithms, and predictive analytics, AI has enabled healthcare providers to identify potential drugs for treatment, optimize treatment protocols, and improve patient outcomes. The integration of AI in these areas not only enhances the efficiency of healthcare services but also supports the ongoing efforts to control and mitigate the impact of the pandemic. In exploring the various applications of AI in COVID-19 treatment and management, this section highlights the innovative strategies and tools that have been developed and their significant impact on public health responses.

7.7.1 Drug Discovery

AI played an essential role in expediting the drug discovery process for COVID-19 treatment. Machine learning algorithms have been utilized to predict the structure of the SARS-CoV-2 virus, thereby identifying potential targets for drug therapy [637, 638]. Furthermore, AI platforms such as DeepMind's AlphaFold have made significant contributions to understanding the protein folding of the virus, which is crucial for the development of antiviral drugs [639]. The deployment of AI in virtual screening has also allowed researchers to rapidly assess millions of chemical compounds, streamlining the identification of viable drug candidates [640, 641].

7.7.2 Patient Management and Monitoring

In the domain of patient management and monitoring, AI systems have been deployed to predict patient outcomes and optimize resource allocation. Predictive analytics have provided healthcare professionals with tools to forecast the progression of the disease in patients, enabling timely interventions [629]. AI-driven algorithms have also been applied to monitor patients' vital signs remotely, reducing the exposure risk for healthcare workers and other patients [642].

7.7.3 Telemedicine

Telemedicine, a component of eHealth, involves using information and communication technology for delivering, managing, and monitoring healthcare services remotely. During the COVID-19 pandemic, telemedicine emerged as a vital tool, especially for patients in isolation [643]. It enabled these patients to receive medical care without the risk of exposing themselves or healthcare providers to the virus. It also alleviated the strain on healthcare facilities, conserved resources like Personal Protective Equipment (PPE), and played a crucial role in the global management of the pandemic.

The surge in demand for healthcare services during the pandemic has underscored the significance of telemedicine, with AI playing a crucial role in its expansion. AI has facilitated remote diagnosis and consultation services, ensuring continuity of care while minimizing the risk of virus transmission [644]. AI-powered chatbots have been employed to provide initial medical assessments based on symptoms reported by patients, thus alleviating the strain on medical facilities [645].

7.8 AI in COVID-19 Prediction and Analytics

Artificial intelligence has also been utilized in the domain of COVID-19 prediction and analytics in the global response to the pandemic. AI models and NLP algorithms have been pivotal in epidemiological modeling, optimizing resource allocation, and analyzing social media to gauge public sentiment and disseminate information.

7.8.1   Epidemiological Modeling

AI has played a critical role in epidemiological modeling, providing forecasts that have been essential for planning and intervention strategies. Sophisticated machine learning models based on reinforcement learning have been employed to predict the spread of the virus, assess the impact of public health interventions, and estimate the burden on healthcare systems [646, 647]. Neural network methods were implemented to identify COVID-19 clusters, providing insights into how socioeconomic factors and spatial distribution are associated with the spread of COVID-19 cases [648]. These models have been crucial in informing government policies, such as lockdowns and vaccination campaigns, to mitigate the spread of the virus [649].

7.8.2   Resource Allocation

In the realm of resource allocation, AI has been instrumental in ensuring the efficient distribution of medical supplies and medical personnel. Predictive analytics have enabled hospitals to anticipate demand for intensive care units (ICU) and ventilators, facilitating timely procurement and allocation of these critical resources [650]. AI has also been used to develop decision-support tools that assist healthcare administrators in making informed decisions about resource distribution, such as determining the need of mechanical ventilation for a COVID-19 patient [651, 652].

7.8.3   Social Media and Sentiment Analysis

AI has been extensively applied to social media platforms to perform sentiment analysis, track misinformation, and understand public perception regarding COVID-19. Natural language processing (NLP) algorithms have analyzed vast amounts of data from social media to identify trends in public discourse, monitor compliance with public health measures, and combat the spread of false information [653]. These insights have been valuable for public health officials to tailor communication strategies and address public concerns effectively [654]. For example, one study in the United States developed an automatic natural language processing pipeline to detect potential COVID-19 cases that might have been untested and unreported from data generated by Twitter users [655].

AI has been an indispensable tool in the fight against COVID-19, offering robust solutions for prediction and analytics. The insights gained from AI applications have not only informed public health strategies but have also played a critical role in managing the social dynamics of the pandemic. As we continue to navigate through these challenging times, AI's's role in prediction and analytics will evolve and become more integrated in multifarious aspects of pandemic response efforts.

7.9 Ethical and Societal Implications

The rapid deployment of AI technologies during the COVID-19 pandemic has given rise to a range of ethical and societal implications that warrant rigorous scrutiny. As AI systems are increasingly integrated into healthcare and public health strategies, concerns surrounding data privacy, algorithmic bias, and accessibility have emerged as critical issues that must be addressed to ensure equitable and ethical use of technology.

7.9.1 Data Privacy

The use of AI in managing the COVID-19 pandemic heavily relies on the collection, processing, and analysis of vast amounts of personal data. Contact tracing apps, health monitoring systems, and AI-driven diagnostic tools all operate on data that is inherently personal and sensitive. The imperative to protect patient confidentiality and adhere to data protection laws is paramount, as breaches can lead to a loss of public trust and potential harm to individuals [656]. The General Data Protection Regulation (GDPR) in the European Union, and similar regulations globally, provide a framework for data protection, but the unprecedented scale of the pandemic poses new challenges in ensuring compliance and safeguarding privacy [657].

7.9.2 Algorithmic Bias

AI algorithms are susceptible to bias, which can arise from skewed training datasets or flawed design and implementation. In the context of COVID-19, such biases can lead to disparities in diagnosis, treatment, and vaccine distribution, disproportionately affecting marginalized communities [658]. It is essential to conduct thorough bias audits and implement corrective measures to mitigate these risks. The development of AI systems must be accompanied by a commitment to the FAIR (Findable, Accessible, Interoperable and Reusable) principles with regard to COVID-19 patient data as well as the inclusion of diverse datasets that reflect the heterogeneity of the population [659].

7.9.3 Accessibility and Inequality

The rapid deployment of AI solutions during the pandemic has highlighted the digital divide and the issue of accessibility. Not all populations have equal access to the technologies that facilitate remote healthcare, such as telemedicine, which can exacerbate existing health inequalities [660]. Furthermore, low-resource settings may lack the infrastructure necessary to implement AI-driven interventions, leading to a disparity in the quality of care and health outcomes [661]. Ensuring equitable access to AI technologies is crucial in the global response to the pandemic and in the broader context of healthcare [662].

The ethical and societal implications of AI in the era of COVID-19 are complex and multifaceted. As we reflect on the challenges posed by the pandemic, it is imperative to foster an ethical AI ecosystem that prioritizes data privacy, mitigates algorithmic bias, and promotes accessibility and equity. Only then can we harness the full potential of AI to serve the greater good without compromising the values of a just and fair society.

7.10 Case Studies

The deployment of AI in response to the COVID-19 pandemic has varied significantly across different countries, with a mix of successes and failures. These case studies provide valuable insights into the potential and limitations of AI in public health emergencies.

7.10.1 Country-Specific Implementations

7.10.1.1 South Korea's AI-Powered Response

South Korea's response to the COVID-19 pandemic is a prime example of effective AI implementation. The country's swift action in developing AI-driven testing, tracing, and treatment strategies resulted in efficient containment of the virus. AI algorithms were employed to analyze travel and medical data, which facilitated rapid contact tracing and targeted testing [663, 664, 665]. Chatbot services such as the Korean COVID-19 chatbot provided citizens with real-time information by integrating public data from the Korea Centers for Disease Control and Prevention (KCDC) and Ministry of Health and Welfare (MOHW) [666], easing the burden on national healthcare hotlines.

7.10.1.2 Singapore's TraceTogether Program

Singapore launched the TraceTogether program, which utilized a mobile application and a token-based system to facilitate digital contact tracing [667]. The technology behind the program assessed the proximity and duration of user interactions to notify individuals of potential exposure to the virus. While the program was innovative, it faced challenges related to user privacy and data security [668]

7.10.1.3   The United States' Vaccine Distribution

In the United States, AI played a crucial role in optimizing vaccine distribution logistics. Recurrent neural networks helped identify optimal locations for vaccine centers and manage supply chains. However, the reliance on AI also led to some disparities in vaccine allocation, highlighting the need for oversight in AI implementations [669].

7.10.2   Success Stories and Failures

AI-driven diagnostic tools have been a success story, with algorithms such as those developed by DeepMind being able to predict the structure of proteins associated with SARS-CoV-2, the virus causing COVID-19 [670]. This breakthrough has implications for understanding the virus's mechanisms and developing treatments.

AI has been successful in disseminating public health messaging via social media platforms, chatbots, and other digital means. These AI systems have been able to tailor messages to specific demographics, improving public engagement and compliance with health guidelines [671].

Conversely, some AI predictive models failed to provide accurate forecasts for the spread of the virus. In many instances, these models were not able to account for the dynamic nature of human behavior and policy changes, leading to over or underestimation of case numbers [672].

These case studies demonstrate that while AI has the potential to significantly aid in pandemic response efforts, its application must be carefully managed to avoid pitfalls. Success depends not only on the technology itself but also on factors such as data quality, user engagement, and the ethical use of AI.

7.11   Future Directions

The COVID-19 pandemic has accelerated the integration of artificial intelligence in healthcare and public health. As we look to the future, there are several emerging technologies and policy recommendations that could shape the next phase of AI in pandemic preparedness and response.

7.11.1 Emerging Technologies

Quantum computing holds the promise of processing complex datasets much faster than traditional computers. In the context of pandemics, quantum algorithms could revolutionize the way we model viral spread, optimize supply chains for medical supplies, and discover new therapeutic drugs [673].

Next-Generation Sequencing (NGS) technologies are rapidly evolving, allowing for quicker and more affordable genomic sequencing. AI, combined with NGS, could enable real-time tracking of pathogen evolution, helping public health officials stay ahead of mutations and variants of concern [674].

Blockchain technology offers a secure and transparent way to manage health data. In pandemics, blockchain can ensure the integrity of health records, facilitate secure data sharing for AI algorithms, and support contact tracing efforts without compromising privacy [675].

7.11.2 Policy Recommendations

Robust data governance frameworks are essential to ensure that AI systems have access to high-quality, representative data while safeguarding individual privacy. Policies must be developed to address data ownership, consent, and anonymization [676].

The global nature of pandemics requires international cooperation. Policy recommendations should encourage the sharing of AI technologies and expertise across borders, as well as collaborative efforts in research and development [677].

To fully harness the potential of AI, investments in education and workforce development are crucial. This includes training healthcare professionals in AI applications and promoting AI literacy among the general population [678].

The future of AI in the context of pandemics is promising, with emerging technologies offering new tools to combat infectious diseases. However, realizing this potential will require thoughtful policy recommendations that promote innovation while addressing ethical, legal, and social implications.

7.12 Conclusion

The COVID-19 pandemic has been a catalyst for unprecedented global change, particularly in the realm of healthcare and technology. Artificial intelligence (AI) has emerged as a critical tool in combating the pandemic, offering solutions for detection, diagnosis, treatment, and management of the disease. It has also played a significant role in understanding and predicting the spread of the virus, aiding in resource allocation, and analyzing public sentiment.

As we reflect on the lessons learned, it is clear that AI has the potential to transform public health responses to future pandemics. However, this potential can only be realized through ethical practices, equitable access, and international collaboration. The integration of AI in healthcare must be approached with a commitment to data privacy, a focus on reducing algorithmic bias, and an emphasis on creating systems that are accessible to all, regardless of socioeconomic status.

The case studies presented throughout this paper highlight both the successes and failures of AI implementations in various contexts, offering valuable insights for future endeavors. Moving forward, emerging technologies such as quantum computing, blockchain, and next-generation sequencing will further enhance the capabilities of AI in public health.

Chapter 8

Intelligence Revolution—The Beginning of the End

8.1 Introduction

8.1.1 Overview of the Intelligence Revolution

Intelligence has been the survivor instinct for humanity, which has empowered humans as a superior species. Throughout evolution, humanity has gone through revolutions, carving the eras that have transformed civilization through advancement.

You can graph human evolution, which is mostly a straight line, but we do get better and change over time, and you can graph technological evolution, which is a line that's going straight up. They are going to intersect each other at some point, and that's happening now [679] - Daniel Wilson With the accelerated adoption of applied Artificial Intelligence (AI), a new era of revolution has sparked the Intelligence Revolution, where the exponential augmentation of Intelligence has accelerated. This augmentation of intelligence was being experimented with, tried, and implemented in the corporate world. Still, the democratization of AI with Generative Pretrained Transformer (GPT), such as Chat GPT from OpenAI, has taken AI's power to the public. This has ignited the velocity of the process of Intelligence Transformation within the Intelligence Revolution.

It's so 2022; this was the discussion starter in our leadership meetings in 2023 when the adoption of ChatGPT shot off the roof. The Intelligence Revolution has begun, which carves a new era of revolution that will transform the way we work, the way the government governs, and the evolution of humans. The first step towards this revolution is being called Intelligence Revolution 1.0 [680].

8.1.1.1 Purpose and Scope of the Chapter

This chapter shall explore the various evolutions that humanity has evolved through for development and modernization, with a comparison and introduction to the newest revolution: the Intelligence Revolution. Further in-depth exploration shall be done as we walk through the journey and transition of the revolution ignited by the adoption and application of AI.

Further, we will also indulge in future casting, where we will explore possible extreme futures from Transhumanism, Civilization Transformation, Corporate Transformation, and Government Transformation through intelligence Transformation under intelligence revolution 1.0. We will also tap into the visionaries' perspectives of the future.

With the current adaptation of AI, 2023 has been the baseline for intelligence transformation through AI democratization; the velocity of transformation, adoption, and growth will also be explored. During this intelligence transformation, human-machine collaboration shall be examined through the Human+Digital+Artificial Intelligence^{TM} Framework [681].

Finally, we will discuss the role of AI in the intelligence revolution and the paradigm shift and evolution of the future of work, architecting the future and, at the same time, highlighting some of the opportunities and technologies that would power the future.

8.2 The Revolutions that Advanced Humanity

The path of humanity spans millennia, an epic story that weaves through the annals of time. From the first stirrings of life to the rich fabric of contemporary society, the evolution of humanity is a tribute to perseverance, inventiveness, and limitless potential.

Human beings have come up with revolutionary ideas and innovations over centuries, together with creative ways to apply those ideas into reality, which is later known as technology. All technologies are born to serve purposes like the man who crafted and designed the earliest human tools (could be an arrow, hammer, or knife). These tools were made purposely to meet his demands like hunting and survival [679].

Forging tools and making fire signaled the beginning of an age distinguished by creativity. Each breakthrough, from agriculture to the printing press, from electric and motorized vehicles to space exploration, and from computing, networking, Internet, and software technologies, moved humanity ahead. It's a saga woven with threads of curiosity and courage, where each era illuminated new vistas of understanding. Let's explore some of the revolutionary eras of humanity.

8.2.1 Pre-Agricultural Era

The Paleolithic Period predates the Agricultural Revolution. It's a pre-agricultural era characterized by early human societies relying on hunting and gathering for sustenance. The Paleolithic Period, predating the Stone Age, witnessed the earliest human tool construction. The Oldowan toolkit, comprising hammer stones, stone cores with flake scars, and sharp stone flakes for cutting [680], emerged during this time. These technological advancements were closely linked to hunting, survival strategies, and food preparation. The discovery of fire, believed to have occurred over 1.3 million years ago in the lower Paleolithic Period, was esteemed by Darwin as one of humanity's most significant discoveries, second only to language [681].

8.2.2 The Era of Agricultural Revolution

The Neolithic Period (12,000–15,000 Years Ago) marked a pivotal era when humans developed crucial technologies such as agriculture and animal farming, laying the groundwork for modern society [682]. Dwellings crafted from mud and clay were constructed, and fabrics fashioned from animal skin and fur became prevalent materials for clothing.

The agricultural era marked a significant shift in human history, transforming human lifestyles and enabling the domestication of plants and animals. This led to a stable food supply, allowing larger populations to thrive in specific regions. This surplus allowed for specialization in skills beyond survival, leading to the emergence of complex societies, technological advancements, and civilizations. The cultivation of crops and the ability to control food production altered human biology, contributing to population growth and physical changes. The proximity to domesticated animals likely led to the evolution of certain diseases but also facilitated immunity and resistance over time. The agricultural era significantly impacted human evolution by facilitating societal development, technological progress, population growth, and alterations in human biology and disease resistance.

8.2.3 Social Revolution—The Transition from Agricultural Society to the Industrial Revolution

The Copper and Bronze Age marked a significant technological advancement, as humans utilized copper and bronze materials to create precise, polished, and shaped mechanical tools through advanced copper smelting technology. Historians and archaeologists have long viewed the discovery of metals and the invention of metallurgy as a revolutionary step in the history of humanity. But metallurgy was more than a technical revolution; its invention in the Bronze Age was primarily a social revolution [683].

The Iron Age introduced iron ore smelting and forging technologies, revolutionizing the production of tools and weapons. Iron-based tools, lighter, stronger, and more costeffective [684] than their bronze and copper counterparts, became the new standard, displacing the use of earlier materials. This transition marked a significant advancement in tool-making capabilities and societal development during this period.

Mesopotamia, the birthplace of civilization, made significant technological advancements, including the discovery of the wheel, early writing systems, arc construction, water transportation, and irrigation techniques. Mesopotamians established the earliest known written language, built the first cities, and advanced in astronomy, mathematics, and art and literature [685].

The Roman Empire, from 25 BC to 390 AD, was a significant period in human history, influencing art, architecture, technology, literature, language, and law. They were known for their engineering expertise, constructing aqueducts, water management systems, and an extensive road network with the famous saying "all roads lead to Rome [686]."

8.2.4   The Industrial Revolution

The Industrial Revolution is the most significant human advancement in the modern technological world, marking the transformation from agricultural revolution and manual manufacturing to machine-augmented manufacturing and production. The augmentation of Physical capabilities was the focus through automation and machine integration. Since the AD 1600s, the world has gone through multiple Industrial Revolutions, which have advanced the world towards modernization, increasing efficiency, quality, and growth. Let's briefly explore the various phases of the Industrial Revolution.

8.2.5   First Industrial Revolution (1600AD Onwards)

The First Industrial Revolution was a watershed moment in human history, altering how we create, construct, and dream, the shift from manual to mechanized production processes. It introduced significant inventions like the utilization of coal for energy, the mechanization of manufacturing, steam-powered locomotives, and the advent of railway transportation. Landmark innovations in communication, such as the electric telegraph (1837) [687], and optical technology, notably the photographic camera by Joseph Nicéphore Niépce (1816) [688], revolutionized various aspects of culture and communication. Key inventions like the automobile (1885–1886) [689] and the telephone (1849) [690] profoundly influenced societal, cultural, and economic perspectives. These inventions have had an everlasting impact on human life, reverberating through nations, cultures, and economies. The value of these inventions resides in their deep, long-lasting influence, a symphony of development that is still shaping our world today.

8.2.6   Second Industrial Revolution (1870AD Onwards)

The Second Industrial Revolution witnessed rapid technological advancements that significantly enhanced the global quality of life. Unlike the macro-inventions of the prior era, this phase relied more on micro-inventions that emphasized productivity and product quality improvements. Notably, it saw a pivotal expansion in electricity usage, enabling larger-scale factories, efficient production, and the development of new communication technologies like the telephone and radio. Steel production increased, synthetic materials emerged, and transportation was revolutionized by automobiles and airplanes.

Two inventions define the Second Industrial Revolution- airplanes and computers. The Wright brothers' 1905 flight of the first airplane [691] marked a significant technological advancement in aviation, allowing humans to fly to unprecedented places. The 1930 invention of the computer [692], initially a calculative device, revolutionized the computer industry with its ability to perform multiple tasks simultaneously. Supercomputers, which can perform complex operations within seconds, have further advanced the aviation industry. Both inventions have significantly impacted modern human civilization.

8.2.7   Third Industrial Revolution (1970AD Onwards)

The Third Industrial Revolution in the 1970s marked a significant shift in manufacturing and automation, integrating memory-programmable controls and early computer systems into industrial processes. This led to a shift towards partial automation, reducing reliance on manual labor. Memory-programmable systems allow for the execution of programmed sequences, enhancing efficiency and productivity in manufacturing processes. Robots, equipped with sensors and sophisticated programming, became instrumental in tasks ranging from assembly line operations in automotive manufacturing to intricate procedures in electronics [693] Further Communication enhancement with the invention of mobile phones, software applications, and the Internet laid a strong foundation for the next industrial revolution.

8.2.8   Fourth Industrial Revolution (2000AD Onwards)

After all this evolution, we are currently in what is famously known as the fourth industrial revolution. It is characterized by increasing automation and the employment of smart machines and smart factories; informed data helps to produce goods more efficiently and productively across the value chain [694].

Industry 4.0 is a new era of interconnected and smart manufacturing, combining information and communication technologies with industrial processes. It is centered on cyber-physical systems, where manufacturing systems are interconnected through networks, forming "smart factories." These factories integrate digital twins, virtual representations of physical assets or processes, allowing for real-time monitoring, analysis, and communication between components and systems. This interconnectedness allows for a level of autonomy in production, allowing machines, components, and personnel to communicate and collaborate seamlessly in the digital world.

Advanced analytics, such as Predictive maintenance, is one of the key advantages of Industry 4.0; this proactive approach minimizes downtime, optimizes efficiency, and reduces maintenance costs [695]. The other notable innovations lie around the adoption of the Internet and the emergence of E-commerce, social media, and the digital world. This is further enhanced by innovations in mobile communication technologies, putting computers in the palms of consumers.

The Fourth Industrial Revolution truly implemented Digital Intelligence, where Computing Power and data storage capacity were rising as per Moore's Law. Expansion of the global network through the Internet was exponential, connecting the world; as of October 2023, there were 5.3 billion Internet users worldwide [682]. Generation of exponential data, where it is estimated that in 2023, about 120 Zetabytes of data was generated; in comparison, in 2010, around 2 Zetabytes were generated, and it is estimated that by 2025, the world will be generating 181 Zetabytes of data [683]. Further innovations around artificial Intelligence and applications of AI models to predict and forecast to increase efficiency and accuracy of the outcome.

8.2.9 *Fifth Industrial Revolution (2020AD Onwards)*

The Fifth Industrial Revolution is evolving as a new phase of industrialization where the emphasis is on human-machine collaboration through advanced technologies such as artificial Intelligence and robotics, laying over the digital transformation of the Fourth Industrial Revolution. This phase of the Industrial Revolution would truly embrace automation and transform the future of work as we see it today, from factory floors to executive board rooms.

As Jeff Bezos predicted, "Because of artificial intelligence and its ability to automate certain tasks that in the past were impossible to automate, not only will we have a much wealthier civilization, but the quality of work will go up very significantly and a higher fraction of people will have callings and careers relative to today [696]" the future of work in collaboration with machines will enhance efficiency, quality, and economic benefits.

With these foundational inventions and adoption, industrial revolutions have paved the way for the next era of revolution beyond the Industrial revolution.

8.2.10   Intelligence Revolution (2023AD Onwards)

The Intelligence Revolution marks a paradigm shift, where the machines would embrace Intelligence, understanding, sensing, evolving, reasoning, and, in some cases, inventing as well. Singularity and the Turing test have been predicted since the 1950s when Alan Turing proposed the Turing test. Now, with advancements in technology and the maturity and adoption of artificial Intelligence, a new era of revolution has emerged; we are calling it the Intelligence Revolution.

8.3   Intelligence Revolution 1.0

As we embrace the Intelligence Revolution 1.0, which is highly focused on Intelligence Transformation. The Intelligence through advanced AI capabilities fueled by Data and computing power generates a new level of Intelligence that machines can generate. There are examples such as ChatGPT, which is based on billions of parameters, reinforcement learning, deep learning capabilities, and generative AI capabilities, which provide human-like responses most intriguingly. Another example can be Autonomous Cars, where Tesla has emerged as a technology company providing car services; there are features such as Full Self-driving capabilities, which allow cars to drive not only on highways but also within city limits, recognizing traffic lights, stop signs, humans, and objects. These technologies have advanced digital & AI technologies such as machine learning, Deep Learning, and Computer Vision, all working in an integrated manner to perform tasks in collaboration with human Intelligence.

Intelligence transformation is the most revolutionary transformation that would impact not only the Economy but also humanity as a whole. With intelligence transformation we will need to learn with intelligent machines, which would displace the majority of the jobs as they exist today and demand new ways of working. An example can be Prompt Generator, a job class that did not exist pre-ChatGPT; such evolution of intelligence calls for strategic planning.

8.3.1   Human-Machine Collaboration: The Mantra to Win with AI in the Era of the Intelligence Revolution

With all the revolutions explained earlier, we just created a background and prepared for what is coming next. Firstly, we would like to define Intelligence into three distinct worlds: Human Intelligence, Digital Intelligence, and Artificial Intelligence; the collaborative approach of these three worlds of Intelligence as formulated by Frost Digital Ventures Human+Digital+Artificial Intelligence^{TM}. This formula of HDA Intelligence^{TM} is architected for intelligence transformation where the power of three intelligence worlds would be harnessed to outperform each of the intelligence worlds performing the task on their own on three levels: efficiency, accuracy and quality.

8.3.2 Human Intelligence

Intelligence is the most powerful asset of Humans. The power of human Intelligence is thinking and feeling before making any decision. There are multiple types of Intelligence that humans carry, which allow humans to make rational decisions, such as Emotional Intelligence, Intuition and Dreams. This would allow human Intelligence to be more creative and innovative.

8.3.3 Digital Intelligence

Digital Intelligence is the Intelligence built within the digital transformation era of the Fourth Industrial Revolution, such as Robotic Process Automation, and Trained AI model automation. Digital Intelligence remains the core foundation for Human Intelligence and Artificial Intelligence interaction.

8.3.4 Artificial Intelligence

AI has always been predicted to be the most transformative technology and is the most important world of the Intelligence Revolution. AI, with the advancement in technologies, is now transforming the intelligence level not only to perform narrow predictions or forecasts but also to autonomous capabilities or acting like a Co-Pilot to augment human Intelligence.

This collaboration of humans and machines will have a significant impact on the intelligence world along with the future of work and will demand workforce transformation.

8.3.5 Future Casting Humanity with Intelligence Revolution–Imaging Beyond 2030

Humanity shall adapt and evolve with the intelligence revolution! How would you imagine the future? Is it going to be like Star Trek with galactic exploration, or is it going to be like exmechanics where there is an Intelligent Robot who is smart and generates emotions, or is it going to be like iRobot where there is mass production of Robots to serve humanity?

Given the infancy of the Intelligence Revolution, we have chosen the future defined by Frost & Sullivan and aligned with the Intelligence Revolution.

8.3.6 Evolution of the Human Body: Transhumanism

Advances in hardware and biological augmentation will enhance physical capabilities. It is defined that human body augmentation would happen through hardware augmentation and biological augmentation, where Hardware augmentation would have the external wearables for sensing enhancements and movement enhancements along with internal implants such as brain implants or body implants. Biological augmentation would be genetic enhancements and genomic and re-productive augmentation. This would enhance human body capabilities such as Enhanced Intelligence, Enhanced sensing, disease-free bodies, sports optimiza-tion, digital hardware enhancements, enhanced strength, increased attractiveness, extreme aging, optimized diet, digital biometrics, and ubiquitous and continual monitoring. With new partners in the Intelligence world enhancing for humans to be competitive and keep pace, a new level of enhancement may be demanded.

8.3.7   Humanity in 2030–Everyday Life

In the intelligence world, civilization may have transformed, which would cre-ate a new world for humanity with enhanced intelligent lifestyles such as BMIs that allow humans to be debugged, insurance tracking diet-driving-lifestyle, in-creased addictive behavior via hyper-personalization, Early Stages of telepathic communication, Disease free humans; nudge practices will evolve towards sus-tainable motivation and manipulation; Knowledge will be freely available; Im-proved quality of life with improved efficiency and productivity; Large portions of leisure time shall be spent in virtual worlds; Greater focus on human values and the drivers of Humanity, Quantification, and optimization of most aspects of life; Intelligence, biology, happiness, and morality compete for human enhance-ment supremacy; improved population behavior related to health, personal debt, and climate change. The intelligent world that humans will create and adapt to shall evolve as part of the everyday lifestyle of humanity.

8.3.8   Humanity in 2030–Corporate World

Given that we are the future casting, humans and lifestyles shall transform, surely that would call for the Corporate world to adapt and evolve as well with intelli-gent workplaces such as AI will be a part of companies' board of directors with company sponsored wearable and nootropics for enhanced productivity; Behav-ioral science driven job applications; Massive retraining efforts to transition of jobs such as truck driver to trip optimizer; Job inequality to be balanced with the emergence of more sage and rewarding jobs; Shift from a hierarchical system to team-based structures [681]. The corporate world shall evolve and thrive by transforming into Intelligent corporations, creating future work, and preparing & demanding the workforce for the future.

8.3.9   Humanity in 2030–Government World

While humanity transforms in aspect towards adapting Intelligence, the govern-ment world shall transform into Intelligent Government by adapting to the world such as Brain Positivity manipulation resulting in fewer wars; State-sponsored actors communicate non-verbally through brain implants and digital technology, population awareness of government nudging leads to transparency mandates; Nation-state sponsored brain race for industry, military and economic power; New Political bodies overseeing AI emergence and related societal transitions; Statesponsored CRISPR super warriors will be in development [681].

As it is prominent that the intelligence revolution would transform humanity to the next level of evolution, this would strongly demand new skills, new jobs, and new intelligence to collaborate with machines.

8.4 The Velocity of Intelligence Revolution: Catching up is not an Option

When we look at eras of revolutions, the pace of transition from one revolution to another has narrowed as technological integration enhanced. The Agricultural Revolution to the First Industrial Revolution took over 10 thousand years. Then, from the First Industrial Revolution to the Second Industrial Revolution, it took almost 300 years; from the Second Industrial Revolution to the Third Industrial Revolution took almost 100 Years; from the Third Industrial Revolution to the Fourth Industrial Revolution took almost 30 years; from the Fourth Industrial Revolution to the Fifth Industrial Revolution took almost 20 Years. Now, the new era of the Intelligence Revolution within three years of the Fifth Industrial Revolution is an exponential leap, and this Intelligence Revolution 1.0 is expected to be transformed into the next era within the next ten years. We saw this transformative velocity in the year 2023, with Open AI breaking all records. Founded in 2015 by prominent leaders such as Elon Musk, Sam Altman, and others, where it released the first consumer version of ChatGPT 3.5 on November 30, 2022, to the public. The velocity of adoption was so high that OpenAI had 1 million users within five days of launch and 100 million users within two months of launch, and by December 2023, OpenAI boasted over 180 million users. OpenAI launched ChatGPT3.5 with over 175 Billion Parameters; by March 2023, ChatGPT 4 with over 1.7 trillion Parameters was released. The Valuation for Open AI was approximately $29bil in April 2023 but had already skyrocketed to around $90billion. If you see the velocity of transformation, we can only imagine what will hold in 2024 with everyone focusing on the Intelligence transformation. We started saying it was in 2022 when old ways of functioning were proposed as 2023 created the age of Intelligence transformation democratized by OpenAI and followed by numerous companies. In 2023, catching up was not an option; you had to adapt to the changing paradigm and evolve, where products were being launched every month, and the world went into a startup mode.

8.4.1   Role of AI in the Intelligence Revolution

"AI will be the best or worst thing ever for humanity" - Elon Musk! It is estimated that A.I has potential to contribute additional $26 trillion to the global economy [688]. With the advancement of AI and accelerated adoption, AI is at the core of the Intelligence Revolution. With advancement of AI, massive data, computing & storage, connected world, large language models; the transformation of intelligence from Human to Machine has been ignited. This would enforce for not only sharing of knowledge and intelligence among colleagues but also to computers and software, for example every prompt that you write in ChatGPT free version, you are training the model what type of prompt is expected and the response to it. Application and adoption of Artificial Intelligence is just beginning, as they say just the tip of iceberg. As we evolve and go deeper into the Intelligence Revolution 1.0: Intelligence Transformation, AI will learn and become more intelligent taking over a majority of the tasks that humans today perform in conjunction with machines such as Autonomous car, Robotic warehouse management, customer experience, Robotic Manufacturing, Research, Digital Twin just to name a few. With this Transformation, Human Intelligence needs evolution towards adapting to work in collaboration with Artificial Intelligence. While the demand for Artificial Intelligence integration both among businesses and consumers is surging, there are fundamental cautions that will need awareness of such as, while generic AI Application is breaking grounds, a narrow AI application which is intelligent and specialized is also necessary, for which there is still a huge shortage of specialized talent in AI Mckinsey highlights that major limiting factors for AI realizing it full business value is an availability of the right skills and capabilities, it is further highlighted that over the survery of 3.5 million jobs posted for AI less than 50% where qualified practitioners. The other aspect of caution is ethical AI, where the governance of AI ensures the right partnership and collaboration between humans and AI in the digital world as intelligence undergoes transformation. International Organization for Standardization (ISO) has started governing the standard AI development and applications such as ISO 24028:2020 for AI Overview of Trustworthiness in AI, ISO 24030:2021 for AI Use Cases, ISO 42001:2023 for AI Management System, ISO 23894:2023 for AI Guidance on Risk Management, ISO 23053:2022 Framework for AI Systems Using Machine Learning to name of few.

Thus, the role of AI in the intelligence revolution is at the core in the age of intelligence revolution. In the future any process would be looked at through the lens of AI first, trickling down to Digital intelligence and then human intelligence. In this scenario each world of intelligence shall perform in collaboration. This would allow humans to focus and excel on human centric intelligence such as Emotional intelligence, Intuition and Visionary innovation to name a few.

8.5 Future of Work: Human Intelligence in the Age of Intelligence Revolution

The workplace as we know is undergoing enormous changes as a result of technological breakthroughs, altering societal expectations, and the worldwide reaction to contemporary concerns.

Imagine a customer service scenario where, instead of navigating endless phone menus, you interact with a friendly AI chatbot that resolves your query in seconds. This is not science fiction; it's a glimpse into the near future, where artificial intelligence takes over the reins of repetitive tasks, freeing up human agents to tackle intricate customer issues that demand empathy, creativity, and critical thinking. For instance, we have already seen the use of AI-driven chatbots in customer support. Routine queries and tasks are handled by these bots, allowing human agents to focus on complex customer issues, thus enhancing the overall quality of service.

In the age of intelligence revolution, The future of work is not just a geographical or technological shift; it represents a fundamental transformation in the essence of work itself. As we look ahead to the future of work, it is clear that the landscape will be defined by innovation, flexibility, and a renewed emphasis on human-centric principles.

8.5.1 Emerging Trends in the Future of Work

With technological advancement and in the age of Intelligence Revolution, future of work looks transformational. It is believed to be larger than the transformation that happened when the Horse Cart was Transformed to Motorized vehicles, where the transformation was feared, but eventually it created a better world, with more jobs and an efficient society. It has already started, with ChatGPT. The new role that is on rise is of Prompt Generator; there is so much more to expect and prepare for. Companies must understand how the future of work will transform due to rising global uncertainties, technological innovations, multigenerational contribution, growing climatic concerns and other factors [680].

8.5.2 Automation to Autonomous: Reimagine Roles

Go forward to a time not too distant from now, when humans and robotics work in perfect harmony to create what is lovingly referred to as "cobots." This period sees automation as a liberator, releasing humans from the shackles of routine, repetitive work, in contrast to anxieties of job displacement. Imagine a workforce in which machines do regular tasks with ease, freeing up humans to express their full creative potential, think strategically, and make difficult decisions. It's about a relationship that changes the fundamental character of work, not about humans against technology.

Intelligent automation will conquer repetitive tasks, which would free up time of the human workforce. Automation would be an intelligent system, that would initiate and complete the task with minimal or no interference from the human workforce. How can we reimagine the roles of three intelligence worlds? human intelligence, digital intelligence and artificial intelligence.

8.5.3 Rise of the Hybrid Workforce and Workplace

Gen Z will account for 32.5% of the future workforce by 2030 [680]. The rise of the Hybrid workforce translates from the following models as per Frost & Sullivan

8.5.3.1 Type of Employment

Hybrid workforce shall encompass a mix of Full Time Employees, Part-time Employees, Leased Employees, Gig Workers and Crowdsourced contributors.

8.5.3.2 Workplace

The hybrid workplace shall encompass muti-model offices such as permanent office location, collocated office, remote locations and fluid offices.

8.5.3.3 Workforce Technology

The hybrid workforce shall encompass collaboration with advanced technologies such as AI, robotics and automation.

The collaborative workforce where AI is within the equation of the workforce and workplace. This shall empower the workforce future in the age of intelligence revolution.

8.5.3.4 Future Brimming with Possibilities: Intelligence Revolution

While the landscape of work is transforming, amidst the evolving trends and challenges, a future brimming with possibilities emerges. A canvas awaits, ready to be painted with vibrant strokes of human ingenuity, technological advancement, and unwavering commitment to progress. Let's dive deeper into these brushstrokes that will shape the future.

8.5.3.5 Continuous Learning: A Lifelong Journey

Gone are the days of monolithic degrees. Micro-credentials - focused skill certifications empower individuals to tailor their learning journeys to specific needs and market demands. These bite-sized knowledge nuggets provide the agility to stay relevant and acquire in-demand skills, opening doors to new opportunities and career advancement. Imagine flexible learning modules tailored to specific needs, accessible anytime, anywhere.

The future workplace becomes a learning incubator. Businesses actively encourage and enable continuous learning through dedicated resources, mentorship programs, and flexible work arrangements. This investment in employee development fosters innovation, adaptability, and a future-proof workforce equipped to navigate any current or future challenge.

Imagine a learning journey personalized just for you. Al-powered platforms act as your guide, recommending relevant courses, identifying knowledge gaps, and ensuring you reach your full potential. This democratization of access to knowledge ensures everyone has the opportunity to become a lifelong learner, continuously refining their skills and reaching their career goals.

8.5.3.6 *Human-AI Collaboration: A Symphony of Strengths*

The fear of Al replacing humans in the workplace misses the mark. The future belongs not to competition, but to human-Al collaboration, a powerful partnership harnessing the unique strengths of both.

The future workplace will see the rise of hybrid teams, seamlessly blending human talent with Al-powered tools. These teams leverage the analytical precision of Al alongside the adaptability and critical thinking of humans, leading to faster and more efficient problem-solving. Picture brainstorming sessions where human intuition sparks new ideas, powered by Al's ability to test and refine those ideas with incredible speed.

Human-Al collaboration isn't a futuristic fantasy; it's the foundation of a thriving future workforce in the age of intelligence revolution. By embracing this partnership, we unlock a world of enhanced productivity, groundbreaking innovation, and shared success.

8.6 Conclusion

This chapter explores the evolution of humanity driving into the age of the Intelligence Revolution, and we go further on how AI has sparked the Intelligence Revolution 1.0. The focus in Intelligence Transformation that is taking place, further focuses on human-machine collaboration as a mantra for success. We then future cast humanity through transhumanism, lifestyle, corporate world and the government world. The velocity of the Intelligence Revolution is exponential, not only in the context of adoption but also production and growth. Where the current model and pace makes 2022 look ancient. Then we focus on how Al is at the core of the Intelligence Revolution. Finally, we dive deeper into the future of work exploring Human Intelligence in the age of Intelligence Revolution. Below are some of the key highlights from this chapter.

8.6.1   Revolutions that has Advanced Humanity

From the Agricultural revolution to the Industrial revolution and embracing into intelligence revolution, where each revolution has one common impact, modernization and in each succeeding revolution the duration of the revolution has diminished substantially.

8.6.2   Intelligence Revolution 1.0

Introduction to Intelligence Transformation, where the intelligence is being transformed into machine level intelligence, thus demanding the understanding of three worlds of intelligence: Human Intelligence, Digital Intelligence and Artificial Intelligence.

8.6.3   Imaging Beyond 2030

The high possibility of AI becoming an integral part of human life is imagined with possibilities.

8.6.4   Focus on the Velocity

Velocity should be the game plan, the pace of transformation is unprecedented.

8.6.5   Role of Artificial Intelligence

Emphasis of AI being the core of the Intelligence Revolution

8.6.6   Future of Workforce and Workplace

Evolution of the role of Human Intelligence in collaboration with AI and how the role of workforce will transform.

Bibliography

[1] C. Metz, K. Weise, N. Grant, and M. Isaac, "Ego, Fear and Money: How the A.I. Fuse Was Lit," *The New York Times*, 2023. https://www.nytimes.com/.

[2] J. McCarthy, M. Minsky, N. Rochester, and C. Shannon, "A Proposal for the Dartmouth Summer Research Project on Artificial Intelligence, August 31, 1955," *AI Mag.*, vol. 27, pp. 12–14, 2006.

[3] M. Guitton, "Artificial vs. enhanced intelligence: Computer or human behavior?," *Comput. Hum. Behav.*, vol. 31, pp. 332–333, 2014.

[4] F. Li and Y. Du, "From AlphaGo to Power System AI: What Engineers Can Learn from Solving the Most Complex Board Game," *IEEE Power and Energy Magazine*, vol. 16, pp. 76–84, 2018.

[5] L.-Q. Shu, Y.-K. Sun, L. Tan, Q. Shu, and A. Chang, "Application of artificial intelligence in pediatrics: past, present and future," *World Journal of Pediatrics*, vol. 15, pp. 105–108, 2019.

[6] H. Geffner, "Artificial Intelligence: From programs to solvers," *AI Commun.*, vol. 27, no. 1, pp. 45–51, 2014.

[7] J. Schmidhuber, "2006: Celebrating 75 Years of AI - History and Outlook: The Next 25 Years," pp. 29–41, 2006.

[8] D. Hassabis, D. Kumaran, C. Summerfield, and M. Botvinick, "Neuroscience-inspired Artificial Intelligence," *Neuron*, vol. 95, no. 2, pp. 245–258, 2017.

[9] D. Silver, A. Huang, C. J. Maddison, A. Guez, L. Sifre, G. van den Driessche, J. Schrittwieser, I. Antonoglou, V. Panneershelvam, M. Lanctot, S. Dieleman, D. Grewe, J. Nham, N. Kalchbrenner, I. Sutskever, T. P.

Lillicrap, M. Leach, K. Kavukcuoglu, T. Graepel, and D. Hassabis, "Mastering the game of Go with deep neural networks and tree search," *Nat.*, vol. 529, no. 7587, pp. 484–489, 2016.

[10] J. Dean, "Google AI: Advancing the State of the Art," *Google AI Blog*, 2018.

[11] L. Floridi, J. Cowls, M. Beltrametti, R. Chatila, P. Chazerand, V. Dignum, C. Luetge, R. Madelin, U. Pagallo, F. Rossi, B. Schafer, P. Valcke, and E. Vayena, "AI4People–An Ethical Framework for a Good AI Society: Opportunities, Risks, Principles, and Recommendations," *Minds and Machines*, vol. 28, no. 4, pp. 689–707, 2018.

[12] D. Silver, T. Hubert, J. Schrittwieser, I. Antonoglou, M. Lai, A. Guez, M. Lanctot, L. Sifre, D. Kumaran, T. Graepel, T. P. Lillicrap, K. Simonyan, and D. Hassabis, "Mastering Chess and Shogi by Self-play with a General Reinforcement Learning Algorithm," *CoRR*, vol. abs/1712.01815, 2017.

[13] G. Brockman, "OpenAI: Advancing the Frontier of Artificial Intelligence," *Harvard Business Review*, 2016.

[14] S. Altman, "OpenAI: Democratizing Artificial Intelligence for the Benefit of Humanity," *TechCrunch*, 2015.

[15] I. Sutskever, "OpenAI: Bridging the Gap Between Research and Implementation," *MIT Technology Review*, 2016.

[16] J. Schulman, "Reinforcement Learning at OpenAI: From Algorithms to Real-world Impact," *Journal of Machine Learning Research*, vol. 18, pp. 1–5, 2017.

[17] T. B. Brown, B. Mann, N. Ryder, M. Subbiah, J. Kaplan, P. Dhariwal, A. Neelakantan, P. Shyam, G. Sastry, A. Askell, S. Agarwal, A. Herbert-Voss, G. Krueger, T. Henighan, R. Child, A. Ramesh, D. M. Ziegler, J. Wu, C. Winter, C. Hesse, M. Chen, E. Sigler, M. Litwin, S. Gray, B. Chess, J. Clark, C. Berner, S. McCandlish, A. Radford, I. Sutskever, and D. Amodei, "Language Models are Few-shot Learners," *CoRR*, vol. abs/2005.14165, 2020.

[18] L. Floridi and M. Chiriatti, "GPT-3: Its Nature, Scope, Limits, and Consequences," *Minds Mach.*, vol. 30, no. 4, pp. 681–694, 2020.

[19] J. Kaplan, S. McCandlish, T. Henighan, T. B. Brown, B. Chess, R. Child, S. Gray, A. Radford, J. Wu, and D. Amodei, "Scaling Laws for Neural Language Models," *CoRR*, vol. abs/2001.08361, 2020.

[20] R. Bommasani, C. Cardie, M. Cranmer, T. Dettmers, I. Sutskever, and P. Liang, "Opportunities and Challenges in the Era of Generative Language Models," *arXiv preprint arXiv:2107.03374*, 2021.

[21] E. M. Bender, T. Gebru, A. McMillan-Major, and S. Shmitchell, "On the Dangers of Stochastic Parrots: Can Language Models Be Too Big?," in *FAccT '21: 2021 ACM Conference on Fairness, Accountability, and Transparency, Virtual Event / Toronto, Canada, March 3-10, 2021* (M. C. Elish, W. Isaac, and R. S. Zemel, eds.), pp. 610–623, ACM, 2021.

[22] K.-F. Lee, *AI Superpowers: China, Silicon Valley, and the New World Order*. Houghton Mifflin Harcourt, 2018.

[23] W. Knight, "Silicon Valley's AI Race," *MIT Technology Review*, 2017.

[24] E. Gibney, "Google DeepMind: What is it, how it works and should you be scared?," *Wired UK*, 2016.

[25] R. Seamans, "AI Startups and Innovation," *AI Magazine*, vol. 38, no. 2, pp. 15–26, 2017.

[26] L. Floridi, *The Fourth Revolution: How the Infosphere is Reshaping Human Reality*. Oxford University Press, 2019.

[27] O. Etzioni, "AI in Silicon Valley: Collaboration and Competition," *AI Magazine*, vol. 37, no. 3, pp. 51–54, 2016.

[28] P. Stone, "AI Talent Wars in Silicon Valley," *AI Magazine*, vol. 37, no. 4, pp. 45–48, 2016.

[29] N. Bostrom, *Superintelligence: Paths, Dangers, Strategies*. Oxford University Press, 2014.

[30] E. Horvitz, "Partnership on AI: Towards a Better Future," *AI Magazine*, vol. 38, no. 4, pp. 90–93, 2017.

[31] S. Russell, *Human Compatible: Artificial Intelligence and the Problem of Control*. Viking, 2019.

[32] M. Abadi, A. Agarwal, P. Barham, E. Brevdo, Z. Chen, C. Citro, G. S. Corrado, A. Davis, J. Dean, M. Devin, *et al.*, "TensorFlow: A system for large-scale machine learning," *12th USENIX Symposium on Operating Systems Design and Implementation (OSDI 16)*, pp. 265–283, 2016.

[33] Y. LeCun, "Facebook AI Research - Advancing the State of the Art," *Facebook AI Research*, 2017.

[34] J. Barr, "Amazon Web Services: Powering Innovation with AI and Machine Learning," *Amazon Web Services Blog*, 2016.

[35] E. Musk, "Why Elon Musk Founded OpenAI," *Business Insider*, 2015.

[36] A. Vance, *Elon Musk: Tesla, SpaceX, and the Quest for a Fantastic Future*. Ecco, 2015.

[37] S. Altman, "OpenAI's Sam Altman on the Future of AI," *Forbes*, 2019.

[38] Y. LeCun, "Facebook AI Research: Building the Future of AI," *Facebook AI Research*, 2015.

[39] S. Zuboff, *The Age of Surveillance Capitalism: The Fight for a Human Future at the New Frontier of Power*. PublicAffairs, 2019.

[40] C. Garvie, A. Bedoya, and J. Frankle, "The Perpetual Line-Up: Unregulated Police Face Recognition in America," *Georgetown Law Center on Privacy & Technology*, 2016.

[41] J. Buolamwini and T. Gebru, "Gender Shades: Intersectional Accuracy Disparities in Commercial Gender Classification," in *Conference on Fairness, Accountability and Transparency, FAT 2018, 23-24 February 2018, New York, NY, USA* (S. A. Friedler and C. Wilson, eds.), vol. 81 of *Proceedings of Machine Learning Research*, pp. 77–91, PMLR, 2018.

[42] C. O'Neil, *Weapons of Math Destruction: How Big Data Increases Inequality and Threatens Democracy*. Crown, 2016.

[43] B. D. Mittelstadt, P. Allo, M. Taddeo, S. Wachter, and L. Floridi, "The ethics of algorithms: Mapping the debate," *Big Data Soc.*, vol. 3, no. 2, p. 205395171667967, 2016.

[44] M. Ford, *Rise of the Robots: Technology and the Threat of a Jobless Future*. Basic Books, 2015.

[45] D. Susskind, *A World Without Work: Technology, Automation, and How We Should Respond*. Metropolitan Books, 2020.

[46] K. Hao, "Here's a list of 58 artificial intelligence companies in healthcare to keep an eye on," *MIT Technology Review*, 2019.

[47] A. Jobin, M. Ienca, and E. Vayena, "The global landscape of AI ethics guidelines," *Nat. Mach. Intell.*, vol. 1, no. 9, pp. 389–399, 2019.

[48] C. B. Frey and M. A. Osborne, *The Future of Employment: How Susceptible Are Jobs to Computerisation?*, vol. 114. Oxford University Press, 2017.

[49] E. Brynjolfsson and A. McAfee, "The Productivity Paradox: Why We're Getting More Innovation but Less Growth," *Foreign Affairs*, vol. 96, no. 5, pp. 148–158, 2017.

[50] J. Mokyr, *Secular Stagnation: The Long View*. National Bureau of Economic Research, 2015.

[51] D. H. Autor, "Why Are There Still So Many Jobs? The History and Future of Workplace Automation," *Journal of Economic Perspectives*, vol. 29, no. 3, pp. 3–30, 2015.

[52] R. Kurzweil, *The Singularity is Near: When Humans Transcend Biology*. Penguin, 2005.

[53] Y. N. Harari, *Homo Deus: A Brief History of Tomorrow*. Harper, 2016.

[54] S. Levy, *In the Plex: How Google Thinks, Works, and Shapes Our Lives*. Simon and Schuster, 2014.

[55] I. Asimov, *I, Robot*. Gnome Press, 1950.

[56] P. K. Dick, *Do Androids Dream of Electric Sheep?* Doubleday, 1968.

[57] S. Kubrick, *2001: A Space Odyssey*. Metro-Goldwyn-Mayer, 1968.

[58]

[59] J. Cellan, *AI in Popular Culture*. Penguin Books, 2015.

[60] A. L. Samuel, "Some Studies in Machine Learning Using the Game of Checkers," *IBM J. Res. Dev.*, vol. 3, no. 3, pp. 210–229, 1959.

[61] Y. LeCun, Y. Bengio, and G. E. Hinton, "Deep learning," *Nat.*, vol. 521, no. 7553, pp. 436–444, 2015.

[62] G. Hinton, L. Deng, D. Yu, G. E. Dahl, A. rahman Mohamed, N. Jaitly, A. Senior, V. Vanhoucke, P. Nguyen, T. N. Sainath, and B. Kingsbury, "Deep Neural Networks for Acoustic Modeling in Speech Recognition," *IEEE Signal Processing Magazine*, vol. 29, no. 6, pp. 82–97, 2012.

[63] S. Thrun, M. Montemerlo, H. Dahlkamp, D. Stavens, A. Aron, J. Diebel, P. W. Fong, J. Gale, M. Halpenny, G. Hoffmann, K. Lau, C. M. Oakley, M. Palatucci, V. R. Pratt, P. Stang, S. Strohband, C. Dupont, L. Jendrossek, C. Koelen, C. Markey, C. Rummel, J. van Niekerk, E. Jensen, P. Alessandrini, G. R. Bradski, B. Davies, S. Ettinger, A. Kaehler, A. V. Nefian, and P. Mahoney, "Stanley: The robot that won the DARPA Grand Challenge," *J. Field Robotics*, vol. 23, no. 9, pp. 661–692, 2006.

[64] A. Esteva, B. Kuprel, R. A. Novoa, J. Ko, S. M. Swetter, H. M. Blau, and S. Thrun, "Dermatologist-level classification of skin cancer with deep neural networks," *Nat.*, vol. 542, no. 7639, pp. 115–118, 2017.

[65] D. J. Trump, "Accelerating America's Leadership in Artificial Intelligence," *The White House*, 2019.

[66] D. A. Ferrucci, A. Levas, S. Bagchi, D. Gondek, and E. T. Mueller, "Watson: Beyond Jeopardy!," *Artif. Intell.*, vol. 199-200, pp. 93–105, 2013.

[67] Y. Bengio, T. Deleu, N. Rahaman, R. N. Ke, S. Lachapelle, O. Bilaniuk, J. Pineau, and A. Courville, "Meta-learning and Universality: Deep Representations and Gradient Descent can Approximate any Learning Algorithm," *arXiv preprint arXiv:1910.11622*, 2019.

[68] J. D. Biamonte, P. Wittek, N. Pancotti, P. Rebentrost, N. Wiebe, and S. Lloyd, "Quantum machine learning," *Nat.*, vol. 549, no. 7671, pp. 195–202, 2017.

[69] X. Zheng, Z. Pang, and W. Sun, "IoT and AI for Smart Healthcare: A Research Bibliography," *IEEE Access*, vol. 6, pp. 24407–24424, 2018.

[70] B. Goertzel, "Artificial General Intelligence," *Scholarpedia*, vol. 10, no. 11, p. 31847, 2015.

[71] C. Metz, "Google's AlphaGo Trounces Humans–But It Also Gives Them a Boost," *Wired*, 2016.

[72] D. Hassabis, D. Silver, A. Huang, and C. J. Maddison, "AlphaGo: Using Machine Learning to Master the Ancient Game of Go," *DeepMind Blog*, 2017.

[73] I. T. Union, "AI for Good Global Summit," 2019.

[74] O. for Economic Co-operation and Development, "OECD Principles on Artificial Intelligence," 2019.

[75] S. Wachter, "Normative challenges of identification in the Internet of Things: Privacy, profiling, discrimination, and the GDPR," *Comput. Law Secur. Rev.*, vol. 34, no. 3, pp. 436–449, 2018.

[76] E. Commission, "Artificial Intelligence Act," 2021.

[77] T. W. House, "Guidance for Regulation of Artificial Intelligence Applications," 2020.

[78] Oct 2021.

[79] C. Smith, "Artificial Intelligence Policies Around the World," *KDnuggets*, 2019.

[80] N. G. Cade Metz, Karen Weise and M. Isaac, "Ego, Fear and Money: How the A.I. Fuse Was Lit," *New York Times*, 2023.

[81] L. E. Baum and T. Petrie, "Statistical inference for probabilistic functions of finite state Markov chains," *The annals of mathematical statistics*, vol. 37, no. 6, pp. 1554–1563, 1966.

[82] L. E. Baum and J. A. Eagon, "An inequality with applications to statistical estimation for probabilistic functions of Markov processes and to a model for ecology," 1967.

[83] L. E. Baum, T. Petrie, G. Soules, and N. Weiss, "A maximization technique occurring in the statistical analysis of probabilistic functions of Markov chains," *The annals of mathematical statistics*, vol. 41, no. 1, pp. 164–171, 1970.

[84] L. R. Rabiner, "A tutorial on hidden Markov models and selected applications in speech recognition," *Proceedings of the IEEE*, vol. 77, no. 2, pp. 257–286, 1989.

[85] J. J. Hopfield, "Neural networks and physical systems with emergent collective computational abilities.," *Proceedings of the national academy of sciences*, vol. 79, no. 8, pp. 2554–2558, 1982.

[86] S. Hochreiter and J. Schmidhuber, "Long Short-term Memory," *Neural Comput.*, vol. 9, no. 8, pp. 1735–1780, 1997.

[87] D. P. Kingma and M. Welling, "An Introduction to Variational Autoencoders," *Found. Trends Mach. Learn.*, vol. 12, no. 4, pp. 307–392, 2019.

[88] D. P. Kingma and M. Welling, "Auto-encoding Variational Bayes," in *2nd International Conference on Learning Representations, ICLR 2014, Banff, AB, Canada, April 14-16, 2014, Conference Track Proceedings* (Y. Bengio and Y. LeCun, eds.), 2014.

[89] A. Creswell, T. White, V. Dumoulin, K. Arulkumaran, B. Sengupta, and A. A. Bharath, "Generative Adversarial Networks: An Overview," *IEEE Signal Process. Mag.*, vol. 35, no. 1, pp. 53–65, 2018.

[90] I. J. Goodfellow, J. Pouget-Abadie, M. Mirza, B. Xu, D. Warde-Farley, S. Ozair, A. C. Courville, and Y. Bengio, "Generative Adversarial Nets," in *Advances in Neural Information Processing Systems 27: Annual Conference on Neural Information Processing Systems 2014, December 8-13 2014, Montreal, Quebec, Canada* (Z. Ghahramani, M. Welling, C. Cortes, N. D. Lawrence, and K. Q. Weinberger, eds.), pp. 2672–2680, 2014.

[91] A. Antoniou, A. J. Storkey, and H. Edwards, "Data Augmentation Generative Adversarial Networks," *CoRR*, vol. abs/1711.04340, 2017.

[92] C. Shorten and T. M. Khoshgoftaar, "A survey on Image Data Augmentation for Deep Learning," *J. Big Data*, vol. 6, p. 60, 2019.

[93] L. Deecke, R. A. Vandermeulen, L. Ruff, S. Mandt, and M. Kloft, "Image Anomaly Detection with Generative Adversarial Networks," in *Machine Learning and Knowledge Discovery in Databases - European Conference, ECML PKDD 2018, Dublin, Ireland, September 10-14, 2018, Proceedings, Part I* (M. Berlingerio, F. Bonchi, T. Gártner, N. Hurley, and G. Ifrim, eds.), vol. 11051 of *Lecture Notes in Computer Science*, pp. 3–17, Springer, 2018.

[94] Q. Yang, P. Yan, Y. Zhang, H. Yu, Y. Shi, X. Mou, M. K. Kalra, Y. Zhang, L. Sun, and G. Wang, "Low-dose CT Image Denoising Using a Generative Adversarial Network With Wasserstein Distance and Perceptual Loss," *IEEE Trans. Medical Imaging*, vol. 37, no. 6, pp. 1348–1357, 2018.

[95] H. Zhang, V. Sindagi, and V. M. Patel, "Image De-raining Using a Conditional Generative Adversarial Network," *IEEE Trans. Circuits Syst. Video Technol.*, vol. 30, no. 11, pp. 3943–3956, 2020.

[96] A. van den Oord, S. Dieleman, H. Zen, K. Simonyan, O. Vinyals, A. Graves, N. Kalchbrenner, A. W. Senior, and K. Kavukcuoglu, "WaveNet: A Generative Model for Raw Audio," in *The 9th ISCA Speech Synthesis Workshop, Sunnyvale, CA, USA, 13-15 September 2016*, p. 125, ISCA, 2016.

[97] A. Ramesh, M. Pavlov, G. Goh, S. Gray, C. Voss, A. Radford, M. Chen, and I. Sutskever, "Zero-shot Text-to-image Generation," in *Proceedings of the 38th International Conference on Machine Learning, ICML 2021, 18-24 July 2021, Virtual Event* (M. Meila and T. Zhang, eds.), vol. 139 of *Proceedings of Machine Learning Research*, pp. 8821–8831, PMLR, 2021.

[98] P. Dhariwal, H. Jun, C. Payne, J. W. Kim, A. Radford, and I. Sutskever, "Jukebox: A Generative Model for Music," *CoRR*, vol. abs/2005.00341, 2020.

[99] E. Cetinic and J. She, "Understanding and Creating Art with AI: Review and Outlook," *ACM Trans. Multim. Comput. Commun. Appl.*, vol. 18, no. 2, pp. 66:1–66:22, 2022.

[100] Y. Bian and X. Xie, "Generative chemistry: drug discovery with deep learning generative models," *CoRR*, vol. abs/2008.09000, 2020.

[101] N. Stephenson, E. Shane, J. Chase, J. Rowland, D. Ries, N. Justice, J. Zhang, L. Chan, and R. Cao, "Survey of machine learning techniques in drug discovery," *Current drug metabolism*, vol. 20, no. 3, pp. 185–193, 2019.

[102] D. Martin, A. Serrano, A. W. Bergman, G. Wetzstein, and B. Masia, "Scangan360: A generative model of realistic scanpaths for 360 images," *IEEE Transactions on Visualization and Computer Graphics*, vol. 28, no. 5, pp. 2003–2013, 2022.

[103] P. Achlioptas, O. Diamanti, I. Mitliagkas, and L. J. Guibas, "Learning Representations and Generative Models for 3D Point Clouds," in *Proceedings of the 35th International Conference on Machine Learning, ICML 2018, Stockholmsmássan, Stockholm, Sweden, July 10-15, 2018* (J. G. Dy and A. Krause, eds.), vol. 80 of *Proceedings of Machine Learning Research*, pp. 40–49, PMLR, 2018.

[104] K. E. Tat, S. P. Lee, A. D. Cheok, S. Kodagoda, Y. Zhou, and G. S. Toh, "Age invaders: social and physical inter-generational family entertainment," in *Extended Abstracts Proceedings of the 2006 Conference on Human Factors in Computing Systems, CHI 2006, Montréal, Québec, Canada, April 22-27, 2006* (G. M. Olson and R. Jeffries, eds.), pp. 243–246, ACM, 2006.

[105] A. Radford, L. Metz, and S. Chintala, "Unsupervised Representation Learning with Deep Convolutional Generative Adversarial Networks," in *4th International Conference on Learning Representations, ICLR 2016, San Juan, Puerto Rico, May 2-4, 2016, Conference Track Proceedings* (Y. Bengio and Y. LeCun, eds.), 2016.

[106] G. P. Way and C. S. Greene, "Extracting a biologically relevant latent space from cancer transcriptomes with variational autoencoders," in *Pacific Symposium on Biocomputing 2018: Proceedings of the Pacific Symposium*, pp. 80–91, World Scientific, 2018.

[107] J. Sirignano and R. Cont, "Universal features of price formation in financial markets: perspectives from deep learning," *Quantitative Finance*, vol. 19, no. 9, pp. 1449–1459, 2019.

[108] M. Reichstein, G. Camps-Valls, B. Stevens, M. Jung, J. Denzler, N. Carvalhais, and Prabhat, "Deep learning and process understanding for data-driven Earth system science," *Nat.*, vol. 566, no. 7743, pp. 195–204, 2019.

[109] A. Vaswani, N. Shazeer, N. Parmar, J. Uszkoreit, L. Jones, A. N. Gomez, L. Kaiser, and I. Polosukhin, "Attention is All you Need," in *Advances in Neural Information Processing Systems 30: Annual Conference on Neural Information Processing Systems 2017, December 4-9, 2017, Long Beach, CA, USA* (I. Guyon, U. von Luxburg, S. Bengio, H. M. Wallach, R. Fergus, S. V. N. Vishwanathan, and R. Garnett, eds.), pp. 5998–6008, 2017.

[110] R. Pascanu, T. Mikolov, and Y. Bengio, "On the difficulty of training recurrent neural networks," in *Proceedings of the 30th International Conference on Machine Learning, ICML 2013, Atlanta, GA, USA, 16-21 June 2013*, vol. 28 of *JMLR Workshop and Conference Proceedings*, pp. 1310–1318, JMLR.org, 2013.

[111] J. Devlin, M. Chang, K. Lee, and K. Toutanova, "BERT: Pre-training of Deep Bidirectional Transformers for Language Understanding," in *Proceedings of the 2019 Conference of the North American Chapter of the Association for Computational Linguistics: Human Language Technologies, NAACL-HLT 2019, Minneapolis, MN, USA, June 2-7, 2019, Volume 1 (Long and Short Papers)* (J. Burstein, C. Doran, and T. Solorio, eds.), pp. 4171–4186, Association for Computational Linguistics, 2019.

[112] A. Rogers, O. Kovaleva, and A. Rumshisky, "A Primer in BERTology: What We Know About How BERT Works," *Trans. Assoc. Comput. Linguistics*, vol. 8, pp. 842–866, 2020.

[113] S. Bubeck, V. Chandrasekaran, R. Eldan, J. Gehrke, E. Horvitz, E. Kamar, P. Lee, Y. T. Lee, Y. Li, S. M. Lundberg, H. Nori, H. Palangi, M. T. Ribeiro, and Y. Zhang, "Sparks of Artificial General Intelligence: Early experiments with GPT-4," *CoRR*, vol. abs/2303.12712, 2023.

[114] W. Jiao, W. Wang, J. Huang, X. Wang, and Z. Tu, "Is ChatGPT A Good Translator? A Preliminary Study," *CoRR*, vol. abs/2301.08745, 2023.

[115] M. Gao, J. Ruan, R. Sun, X. Yin, S. Yang, and X. Wan, "Human-like Summarization Evaluation with ChatGPT," *CoRR*, vol. abs/2304.02554, 2023.

[116] A. Dosovitskiy, L. Beyer, A. Kolesnikov, D. Weissenborn, X. Zhai, T. Unterthiner, M. Dehghani, M. Minderer, G. Heigold, S. Gelly, J. Uszkoreit, and N. Houlsby, "An Image is Worth 16x16 Words: Transformers for Image Recognition at Scale," in *9th International Conference on Learning Representations, ICLR 2021, Virtual Event, Austria, May 3-7, 2021*, OpenReview.net, 2021.

[117] M. Raghu, T. Unterthiner, S. Kornblith, C. Zhang, and A. Dosovitskiy, "Do Vision Transformers See Like Convolutional Neural Networks?," in *Advances in Neural Information Processing Systems 34: Annual Conference on Neural Information Processing Systems 2021, NeurIPS 2021, December 6-14, 2021, virtual* (M. Ranzato, A. Beygelzimer, Y. N. Dauphin, P. Liang, and J. W. Vaughan, eds.), pp. 12116–12128, 2021.

[118] S. Paul and P. Chen, "Vision Transformers Are Robust Learners," in *Thirty-Sixth AAAI Conference on Artificial Intelligence, AAAI 2022,*

Thirty-Fourth Conference on Innovative Applications of Artificial Intelligence, IAAI 2022, The Twelveth Symposium on Educational Advances in Artificial Intelligence, EAAI 2022 Virtual Event, February 22 - March 1, 2022, pp. 2071–2081, AAAI Press, 2022.

[119] G. S. Nikolić, B. R. Dimitrijević, T. R. Nikolić, and M. K. Stojcev, "A survey of three types of processing units: CPU, GPU and TPU," in *2022 57th International Scientific Conference on Information, Communication and Energy Systems and Technologies (ICEST)*, pp. 1–6, IEEE, 2022.

[120] R. Gozalo-Brizuela and E. C. Garrido-Merchán, "ChatGPT is not all you need. A State of the Art Review of large Generative AI models," *CoRR*, vol. abs/2301.04655, 2023.

[121] T. Lin, Y. Wang, X. Liu, and X. Qiu, "A Survey of Transformers," *CoRR*, vol. abs/2106.04554, 2021.

[122] K. S. Kalyan, A. Rajasekharan, and S. Sangeetha, "AMMUS : A Survey of Transformer-based Pretrained Models in Natural Language Processing," *CoRR*, vol. abs/2108.05542, 2021.

[123] F. A. Acheampong, H. Nunoo-Mensah, and W. Chen, "Transformer models for text-based emotion detection: a review of BERT-based approaches," *Artif. Intell. Rev.*, vol. 54, no. 8, pp. 5789–5829, 2021.

[124] K. Han, Y. Wang, H. Chen, X. Chen, J. Guo, Z. Liu, Y. Tang, A. Xiao, C. Xu, Y. Xu, Z. Yang, Y. Zhang, and D. Tao, "A Survey on Vision Transformer," *IEEE Trans. Pattern Anal. Mach. Intell.*, vol. 45, no. 1, pp. 87–110, 2023.

[125] S. H. Khan, M. Naseer, M. Hayat, S. W. Zamir, F. S. Khan, and M. Shah, "Transformers in Vision: A Survey," *ACM Comput. Surv.*, vol. 54, no. 10s, pp. 200:1–200:41, 2022.

[126] F. Shamshad, S. Khan, S. W. Zamir, M. H. Khan, M. Hayat, F. S. Khan, and H. Fu, "Transformers in Medical Imaging: A Survey," *CoRR*, vol. abs/2201.09873, 2022.

[127] A. A. Aleissaee, A. Kumar, R. M. Anwer, S. Khan, H. Cholakkal, G. Xia, and F. S. Khan, "Transformers in Remote Sensing: A Survey," *Remote. Sens.*, vol. 15, no. 7, p. 1860, 2023.

[128] Q. Wen, T. Zhou, C. Zhang, W. Chen, Z. Ma, J. Yan, and L. Sun, "Transformers in Time Series: A Survey," in *Proceedings of the Thirty-Second International Joint Conference on Artificial Intelligence, IJCAI 2023, 19th-25th August 2023, Macao, SAR, China*, pp. 6778–6786, ijcai.org, 2023.

[129] S. Ahmed, I. E. Nielsen, A. Tripathi, S. Siddiqui, R. P. Ramachandran, and G. Rasool, "Transformers in Time-series Analysis: A Tutorial," *Circuits Syst. Signal Process.*, vol. 42, no. 12, pp. 7433–7466, 2023.

[130] A. M. Turing *et al.*, "On computable numbers, with an application to the Entscheidungsproblem," *J. of Math*, vol. 58, no. 345-363, p. 5, 1936.

[131] B. J. Copeland, "The church-turing thesis," 1997.

[132] P. Bernays, "Alonzo Church. An unsolvable problem of elementary number theory. American journal of mathematics, vol. 58 (1936), pp. 345–363.," *The Journal of Symbolic Logic*, vol. 1, no. 2, pp. 73–74, 1936.

[133] A. Hodges, *Alan Turing: The Enigma: The Book That Inspired the Film The Imitation Game*. Princeton University Press, 2014.

[134] A. M. Turing *et al.*, "Proposed electronic calculator," *National Physical Laboratory*, 1946.

[135] C. Machinery, "Computing machinery and intelligence-AM Turing," *Mind*, vol. 59, no. 236, p. 433, 1950.

[136] A. Turing, "Intelligent machinery (1948)," *B. Jack Copeland*, p. 395, 2004.

[137] A. M. Turing, "The chemical basis of morphogenesis," *Philosophical Transactions of the Royal Society of London. Series B, Biological Sciences*, vol. 237, no. 641, pp. 37–72, 1952.

[138] I. J. Goodfellow, Y. Bengio, and A. C. Courville, *Deep Learning*. Adaptive computation and machine learning, MIT Press, 2016.

[139] C. M. Bishop and N. M. Nasrabadi, *Pattern recognition and machine learning*, vol. 4. Springer, 2006.

[140] Y. Yu, X. Si, C. Hu, and J. Zhang, "A Review of Recurrent Neural Networks: LSTM Cells and Network Architectures," *Neural Comput.*, vol. 31, no. 7, pp. 1235–1270, 2019.

[141] A. Radford, K. Narasimhan, T. Salimans, I. Sutskever, *et al.*, "Improving language understanding by generative pre-training," 2018.

[142] A. Radford, J. Wu, R. Child, D. Luan, D. Amodei, I. Sutskever, *et al.*, "Language models are unsupervised multitask learners," *OpenAI blog*, vol. 1, no. 8, p. 9, 2019.

[143] OpenAI, "GPT-4 Technical Report," *CoRR*, vol. abs/2303.08774, 2023.

[144] H. Touvron, T. Lavril, G. Izacard, X. Martinet, M. Lachaux, T. Lacroix, B. Rozière, N. Goyal, E. Hambro, F. Azhar, A. Rodriguez, A. Joulin, E. Grave, and G. Lample, "LLaMA: Open and Efficient Foundation Language Models," *CoRR*, vol. abs/2302.13971, 2023.

[145] R. Thoppilan, D. D. Freitas, J. Hall, N. Shazeer, A. Kulshreshtha, H. Cheng, A. Jin, T. Bos, L. Baker, Y. Du, Y. Li, H. Lee, H. S. Zheng, A. Ghafouri, M. Menegali, Y. Huang, M. Krikun, D. Lepikhin, J. Qin, D. Chen, Y. Xu, Z. Chen, A. Roberts, M. Bosma, Y. Zhou, C. Chang, I. Krivokon, W. Rusch, M. Pickett, K. S. Meier-Hellstern, M. R. Morris, T. Doshi, R. D. Santos, T. Duke, J. Soraker, B. Zevenbergen, V. Prabhakaran, M. Diaz, B. Hutchinson, K. Olson, A. Molina, E. Hoffman-John, J. Lee, L. Aroyo, R. Rajakumar, A. Butryna, M. Lamm, V. Kuzmina, J. Fenton, A. Cohen, R. Bernstein, R. Kurzweil, B. A. y Arcas, C. Cui, M. Croak, E. H. Chi, and Q. Le, "LaMDA: Language Models for Dialog Applications," *CoRR*, vol. abs/2201.08239, 2022.

[146] B. Zhuang, J. Liu, Z. Pan, H. He, Y. Weng, and C. Shen, "A Survey on Efficient Training of Transformers," in *Proceedings of the Thirty-Second International Joint Conference on Artificial Intelligence, IJCAI 2023, 19th-25th August 2023, Macao, SAR, China*, pp. 6823–6831, ijcai.org, 2023.

[147] F. F. Xu, U. Alon, G. Neubig, and V. J. Hellendoorn, "A Systematic Evaluation of Large Language Models of Code," *CoRR*, vol. abs/2202.13169, 2022.

[148] J. Hewitt, C. D. Manning, and P. Liang, "Truncation Sampling as Language Model Desmoothing," in *Findings of the Association for Computational Linguistics: EMNLP 2022, Abu Dhabi, United Arab Emirates, December 7-11, 2022* (Y. Goldberg, Z. Kozareva, and Y. Zhang, eds.), pp. 3414–3427, Association for Computational Linguistics, 2022.

[149] J. Zhang, T. He, S. Sra, and A. Jadbabaie, "Why Gradient Clipping Accelerates Training: A Theoretical Justification for Adaptivity," in *8th International Conference on Learning Representations, ICLR 2020, Addis Ababa, Ethiopia, April 26-30, 2020*, OpenReview.net, 2020.

[150] Y. Lin, S. Han, H. Mao, Y. Wang, and B. Dally, "Deep Gradient Compression: Reducing the Communication Bandwidth for Distributed Training," in *6th International Conference on Learning Representations, ICLR 2018, Vancouver, BC, Canada, April 30 - May 3, 2018, Conference Track Proceedings*, OpenReview.net, 2018.

[151] M. Shoeybi, M. Patwary, R. Puri, P. LeGresley, J. Casper, and B. Catanzaro, "Megatron-LM: Training Multi-billion Parameter Language Models Using Model Parallelism," *CoRR*, vol. abs/1909.08053, 2019.

[152] D. M. Ziegler, N. Stiennon, J. Wu, T. B. Brown, A. Radford, D. Amodei, P. F. Christiano, and G. Irving, "Fine-tuning Language Models from Human Preferences," *CoRR*, vol. abs/1909.08593, 2019.

[153] J. Dodge, G. Ilharco, R. Schwartz, A. Farhadi, H. Hajishirzi, and N. A. Smith, "Fine-tuning Pretrained Language Models: Weight Initializations, Data Orders, and Early Stopping," *CoRR*, vol. abs/2002.06305, 2020.

[154] R. He, L. Liu, H. Ye, Q. Tan, B. Ding, L. Cheng, J. Low, L. Bing, and L. Si, "On the Effectiveness of Adapter-based Tuning for Pretrained Language Model Adaptation," in *Proceedings of the 59th Annual Meeting of the Association for Computational Linguistics and the 11th International Joint Conference on Natural Language Processing, ACL/IJCNLP 2021, (Volume 1: Long Papers), Virtual Event, August 1-6, 2021* (C. Zong, F. Xia, W. Li, and R. Navigli, eds.), pp. 2208–2222, Association for Computational Linguistics, 2021.

[155] M. Shidiq, "The use of artificial intelligence-based chat-gpt and its challenges for the world of education; from the viewpoint of the development of creative writing skills," in *Proceeding of International Conference on Education, Society and Humanity*, vol. 1, pp. 353–357, 2023.

[156] D. Ippolito, A. Yuan, A. Coenen, and S. Burnam, "Creative Writing with an AI-powered Writing Assistant: Perspectives from Professional Writers," *CoRR*, vol. abs/2211.05030, 2022.

[157] N. Kóbis and L. D. Mossink, "Artificial intelligence versus Maya Angelou: Experimental evidence that people cannot differentiate AI-generated from human-written poetry," *Computers in human behavior*, vol. 114, p. 106553, 2021.

[158] M. Hardalov, I. Koychev, and P. Nakov, "Towards Automated Customer Support," in *Artificial Intelligence: Methodology, Systems, and Applications - 18th International Conference, AIMSA 2018, Varna, Bulgaria, September 12-14, 2018, Proceedings* (G. Agre, J. van Genabith, and T. Declerck, eds.), vol. 11089 of *Lecture Notes in Computer Science*, pp. 48–59, Springer, 2018.

[159] A. Følstad and M. Skjuve, "Chatbots for customer service: user experience and motivation," in *Proceedings of the 1st International Conference on Conversational User Interfaces, CUI 2019, Dublin, Ireland, August 22-23, 2019* (B. R. Cowan and L. Clark, eds.), pp. 1:1–1:9, ACM, 2019.

[160] J. Finnie-Ansley, P. Denny, B. A. Becker, A. Luxton-Reilly, and J. Prather, "The Robots Are Coming: Exploring the Implications of OpenAI Codex

on Introductory Programming," in *ACE '22: Australasian Computing Education Conference, Virtual Event, Australia, February 14 - 18, 2022* (J. Sheard and P. Denny, eds.), pp. 10–19, ACM, 2022.

[161] S. Vártinen, P. Hámálbias' in nlpainen, and C. Guckelsberger, "Generating role-playing game quests with gpt language models," *IEEE Transactions on Games*, 2022.

[162] F. Doshi-Velez and B. Kim, "Towards a rigorous science of interpretable machine learning," *arXiv preprint arXiv:1702.08608*, 2017.

[163] K. Xu, J. Ba, R. Kiros, K. Cho, A. C. Courville, R. Salakhutdinov, R. S. Zemel, and Y. Bengio, "Show, Attend and Tell: Neural Image Caption Generation with Visual Attention," in *Proceedings of the 32nd International Conference on Machine Learning, ICML 2015, Lille, France, 6-11 July 2015* (F. R. Bach and D. M. Blei, eds.), vol. 37 of *JMLR Workshop and Conference Proceedings*, pp. 2048–2057, JMLR.org, 2015.

[164] H. Chefer, S. Gur, and L. Wolf, "Transformer Interpretability Beyond Attention Visualization," in *IEEE Conference on Computer Vision and Pattern Recognition, CVPR 2021, virtual, June 19-25, 2021*, pp. 782–791, Computer Vision Foundation / IEEE, 2021.

[165] N. Elhage, N. Nanda, C. Olsson, T. Henighan, N. Joseph, B. Mann, A. Askell, Y. Bai, A. Chen, T. Conerly, *et al.*, "A mathematical framework for transformer circuits," *Transformer Circuits Thread*, vol. 1, 2021.

[166] Z. Ji, N. Lee, R. Frieske, T. Yu, D. Su, Y. Xu, E. Ishii, Y. Bang, A. Madotto, and P. Fung, "Survey of Hallucination in Natural Language Generation," *ACM Comput. Surv.*, vol. 55, no. 12, pp. 248:1–248:38, 2023.

[167] D. Ganguli, D. Hernandez, L. Lovitt, A. Askell, Y. Bai, A. Chen, T. Conerly, N. DasSarma, D. Drain, N. Elhage, S. E. Showk, S. Fort, Z. Hatfield-Dodds, T. Henighan, S. Johnston, A. Jones, N. Joseph, J. Kernian, S. Kravec, B. Mann, N. Nanda, K. Ndousse, C. Olsson, D. Amodei, T. Brown, J. Kaplan, S. McCandlish, C. Olah, D. Amodei, and J. Clark, "Predictability and Surprise in Large Generative Models," in *FAccT '22: 2022 ACM Conference on Fairness, Accountability, and Transparency, Seoul, Republic of Korea, June 21 - 24, 2022*, pp. 1747–1764, ACM, 2022.

[168] A. Silva, P. Tambwekar, and M. C. Gombolay, "Towards a Comprehensive Understanding and Accurate Evaluation of Societal Biases in Pretrained Transformers," in *Proceedings of the 2021 Conference of the North American Chapter of the Association for Computational Linguistics: Human Language Technologies, NAACL-HLT 2021, Online, June 6-11, 2021* (K. Toutanova, A. Rumshisky, L. Zettlemoyer, D. Hakkani-Túr, I. Beltagy,

S. Bethard, R. Cotterell, T. Chakraborty, and Y. Zhou, eds.), pp. 2383–2389, Association for Computational Linguistics, 2021.

[169] C. Li, "OpenAI's GPT-3 Language Model: A Technical Overview," *Lambda Labs Blog*, 2020. Accessed: 23 October 2023.

[170] P. Xu, X. Zhu, and D. A. Clifton, "Multimodal Learning With Transformers: A Survey," *IEEE Trans. Pattern Anal. Mach. Intell.*, vol. 45, no. 10, pp. 12113–12132, 2023.

[171] S. Pal, M. Bhattacharya, S.-S. Lee, and C. Chakraborty, "A Domain-specific Next-generation Large Language Model (LLM) or ChatGPT is Required for Biomedical Engineering and Research," *Annals of Biomedical Engineering*, pp. 1–4, 2023.

[172] C. Wang, X. Liu, Y. Yue, X. Tang, T. Zhang, J. Cheng, Y. Yao, W. Gao, X. Hu, Z. Qi, Y. Wang, L. Yang, J. Wang, X. Xie, Z. Zhang, and Y. Zhang, "Survey on Factuality in Large Language Models: Knowledge, Retrieval and Domain-specificity," *CoRR*, vol. abs/2310.07521, 2023.

[173] S. Wu, O. Irsoy, S. Lu, V. Dabravolski, M. Dredze, S. Gehrmann, P. Kambadur, D. S. Rosenberg, and G. Mann, "BloombergGPT: A Large Language Model for Finance," *CoRR*, vol. abs/2303.17564, 2023.

[174] L. Floridi, J. Cowls, M. Beltrametti, R. Chatila, P. Chazerand, V. Dignum, C. Luetge, R. Madelin, U. Pagallo, F. Rossi, *et al.*, "An ethical framework for a good AI society: Opportunities, risks, principles, and recommendations," *Ethics, governance, and policies in artificial intelligence*, pp. 19–39, 2021.

[175] F. A. Raso, H. Hilligoss, V. Krishnamurthy, C. Bavitz, and L. Kim, "Artificial intelligence & human rights: Opportunities & risks," 2018.

[176] S. J. Russell and P. Norvig, *Artificial intelligence: a modern approach.* Pearson, 2021.

[177] T. Lee and H. Lee, *Credit scoring models with intelligence techniques.* Springer, 2019.

[178] M. I. Jordan and T. M. Mitchell, "Machine learning: Trends, perspectives, and prospects," *Science*, vol. 349, no. 6245, pp. 255–260, 2015.

[179] R. J. Bolton and D. J. Hand, "Statistical fraud detection: A review," *Statistical science*, pp. 235–249, 2002.

[180] E. J. Topol, "High-performance medicine: the convergence of human and artificial intelligence," *Nature Medicine*, vol. 25, no. 1, pp. 44–56, 2019.

[181] C. Rudin, "Stop explaining black box machine learning models for high stakes decisions and use interpretable models instead," *Nat. Mach. Intell.*, vol. 1, no. 5, pp. 206–215, 2019.

[182] A. Babuta, *Big data and policing: An assessment of law enforcement requirements, expectations and priorities.* Rand Corporation, 2017.

[183] I. Ajunwa, "The paradox of automation as anti-bias intervention," *Cardozo L. Rev.*, vol. 41, p. 1671, 2020.

[184] N. Mehrabi, F. Morstatter, N. Saxena, K. Lerman, and A. Galstyan, "A Survey on Bias and Fairness in Machine Learning," *ACM Comput. Surv.*, vol. 54, no. 6, pp. 115:1–115:35, 2022.

[185] T. Bolukbasi, K. Chang, J. Y. Zou, V. Saligrama, and A. T. Kalai, "Man is to Computer Programmer as Woman is to Homemaker? Debiasing Word Embeddings," in *Advances in Neural Information Processing Systems 29: Annual Conference on Neural Information Processing Systems 2016, December 5-10, 2016, Barcelona, Spain* (D. D. Lee, M. Sugiyama, U. von Luxburg, I. Guyon, and R. Garnett, eds.), pp. 4349–4357, 2016.

[186] R. Benjamin, *Race after technology: Abolitionist tools for the new jim code.* John Wiley & Sons, 2019.

[187] M. Fourcade and K. Healy, "Seeing like a market," *Socio-Economic Review*, vol. 15, no. 1, pp. 9–29, 2017.

[188] M. C. Elish and D. Boyd, "Situating methods in the magic of Big Data and AI," *Communication Monographs*, vol. 85, no. 1, pp. 57–80, 2018.

[189] B. Lepri, N. Oliver, E. Letouzé, A. Pentland, and P. Vinck, "Fair, transparent, and accountable algorithmic decision-making processes," *Philosophy & Technology*, vol. 31, no. 4, pp. 611–627, 2018.

[190] S. Wachter, B. Mittelstadt, and L. Floridi, "Transparent, explainable, and accountable AI for robotics," *Science Robotics*, vol. 2, no. 6, 2017.

[191] J. Kleinberg, J. Ludwig, S. Mullainathan, and A. Rambachan, "Algorithmic fairness," *AEA Papers and Proceedings*, vol. 108, pp. 22–27, 2018.

[192] J. Angwin, J. Larson, S. Mattu, and L. Kirchner, "Machine bias," *ProPublica*, vol. 23, 2016.

[193] R. Dotan and S. Milli, "Value-laden disciplinary shifts in machine learning," in *Proceedings of the AAAI/ACM Conference on AI, Ethics, and Society*, pp. 353–359, 2019.

[194] A. Z. Jacobs and H. Wallach, "Measurement and fairness," in *Proceedings of the 2021 ACM Conference on Fairness, Accountability, and Transparency*, pp. 375–385, 2021.

[195] L. Oakden-Rayner, J. Dunnmon, G. Carneiro, and C. Ré, "Hidden stratification causes clinically meaningful failures in machine learning for medical imaging," in *ACM CHIL '20: ACM Conference on Health, Inference, and Learning, Toronto, Ontario, Canada, April 2-4, 2020 [delayed]* (M. Ghassemi, ed.), pp. 151–159, ACM, 2020.

[196] M. Bickert, C. Scheele, and S. Theves, "Women in AI Research and Practice: Chances and Challenges," *KI-Künstliche Intelligenz*, vol. 35, no. 2, pp. 235–239, 2021.

[197] I. D. Raji, A. Smart, R. N. White, M. Mitchell, T. Gebru, B. Hutchinson, J. Smith-Loud, D. Theron, and P. Barnes, "Closing the AI Accountability Gap: Defining an End-to-end Framework for Internal Algorithmic Auditing," *CoRR*, vol. abs/2001.00973, pp. 33–44, 2020.

[198] Z. C. Lipton, "The mythos of model interpretability: In machine learning, the concept of interpretability is both important and slippery," *Queue*, vol. 16, no. 3, pp. 31–57, 2018.

[199] A. D. Selbst, danah boyd, S. A. Friedler, S. Venkatasubramanian, and J. Vertesi, "Fairness and Abstraction in Sociotechnical Systems," in *Proceedings of the Conference on Fairness, Accountability, and Transparency, FAT* 2019, Atlanta, GA, USA, January 29-31, 2019* (danah boyd and J. H. Morgenstern, eds.), pp. 59–68, ACM, 2019.

[200] A. Birhane, "Algorithmic injustice: a relational ethics approach," *Patterns*, vol. 2, no. 2, p. 100205, 2021.

[201] S. Mitchell, E. Potash, S. Barocas, A. D'Amour, and K. Lum, "Algorithmic fairness: Choices, assumptions, and definitions," *Annual Review of Statistics and Its Application*, vol. 8, pp. 141–163, 2021.

[202] K. Holstein, J. W. Vaughan, H. D. III, M. Dudík, and H. M. Wallach, "Improving Fairness in Machine Learning Systems: What Do Industry Practitioners Need?," in *Proceedings of the 2019 CHI Conference on Human Factors in Computing Systems, CHI 2019, Glasgow, Scotland, UK, May 04-09, 2019* (S. A. Brewster, G. Fitzpatrick, A. L. Cox, and V. Kostakos, eds.), p. 600, ACM, 2019.

[203] Z. Obermeyer, B. Powers, C. Vogeli, and S. Mullainathan, "Dissecting racial bias in an algorithm used to manage the health of populations," *Science*, vol. 366, no. 6464, pp. 447–453, 2019.

[204] J. Dastin, "Amazon scraps secret AI recruiting tool that showed bias against women," *Reuters*, Oct 2018.

[205] Y. Feinstein, "Solving the AI Bias Problem Using Causal Reasoning," *arXiv preprint arXiv:2002.09186*, 2020.

[206] B. Green, *The Smart Enough City: Putting Technology In Its Place To Reclaim Our Urban Future*. MIT Press, 2020.

[207] Z. Tufekci, "Algorithmic harms beyond Facebook and Google: Emergent challenges of computational agency," *Colorado Technology Law Journal*, vol. 13, no. 2, pp. 203–218, 2015.

[208] H. Suresh and J. V. Guttag, "A Framework for Understanding Unintended Consequences of Machine Learning," *CoRR*, vol. abs/1901.10002, 2019.

[209] M. A. Katell, M. Young, D. Dailey, B. Herman, V. Guetler, A. Tam, C. Bintz, D. Raz, and P. M. Krafft, "Toward situated interventions for algorithmic equity: lessons from the field," in *FAT* '20: Conference on Fairness, Accountability, and Transparency, Barcelona, Spain, January 27-30, 2020* (M. Hildebrandt, C. Castillo, L. E. Celis, S. Ruggieri, L. Taylor, and G. Zanfir-Fortuna, eds.), pp. 45–55, ACM, 2020.

[210] D. Zhang, N. Maslej, E. Brynjolfsson, J. Etchemendy, T. Lyons, J. Manyika, H. Ngo, J. C. Niebles, M. Sellitto, E. Sakhaee, Y. Shoham, J. Clark, and C. R. Perrault, "The AI Index 2022 Annual Report," *CoRR*, vol. abs/2205.03468, 2022.

[211] J. Sanchez-Monedero, L. Dencik, and L. Edwards, "What does it mean to 'solve' discrimination in hiring? Social, technical and legal perspectives from the UK on automated hiring systems," in *Proceedings of the 2020 Conference on Fairness, Accountability, and Transparency*, pp. 458–468, 2020.

[212] T. Hagendorff, "The Ethics of AI Ethics: An Evaluation of Guidelines," *Minds Mach.*, vol. 30, no. 1, pp. 99–120, 2020.

[213] K. Crawford, *The atlas of AI*. Yale University Press, 2021.

[214] J. Rawls, *A theory of justice*. Harvard University Press, 1971.

[215] B. Mittelstadt, "Principles alone cannot guarantee ethical AI," *Nature Machine Intelligence*, vol. 1, no. 11, pp. 501–507, 2019.

[216] S. A. Friedler, C. Scheidegger, S. Venkatasubramanian, S. Choudhary, E. P. Hamilton, and D. Roth, "A comparative study of fairness-enhancing interventions in machine learning," in *Proceedings of the Conference on Fairness, Accountability, and Transparency, FAT* 2019, Atlanta, GA,*

USA, January 29-31, 2019 (danah boyd and J. H. Morgenstern, eds.), pp. 329–338, ACM, 2019.

[217] S. L. Blodgett, S. Barocas, H. D. III, and H. M. Wallach, "Language (Technology) is Power: A Critical Survey of "Bias" in NLP," in *Proceedings of the 58th Annual Meeting of the Association for Computational Linguistics, ACL 2020, Online, July 5-10, 2020* (D. Jurafsky, J. Chai, N. Schluter, and J. R. Tetreault, eds.), pp. 5454–5476, Association for Computational Linguistics, 2020.

[218] A. D. Selbst and S. Barocas, "The intuitive appeal of explainable machines," *Fordham L. Rev.*, vol. 87, p. 1085, 2018.

[219] S. K. Katyal, "Private accountability in the age of artificial intelligence," *UCLA L. Rev.*, vol. 66, p. 54, 2019.

[220] J. Whittlestone, R. Nyrup, A. Alexandrova, and S. Cave, "The Role and Limits of Principles in AI Ethics: Towards a Focus on Tensions," in *Proceedings of the 2019 AAAI/ACM Conference on AI, Ethics, and Society, AIES 2019, Honolulu, HI, USA, January 27-28, 2019* (V. Conitzer, G. K. Hadfield, and S. Vallor, eds.), pp. 195–200, ACM, 2019.

[221] A. L. Hoffmann, "Where fairness fails: Data, algorithms, and the limits of anti-discrimination discourse," *Information, Communication & Society*, vol. 22, no. 7, pp. 900–915, 2019.

[222] S. D. Warren and L. D. Brandeis, "The right to privacy," *Harvard Law Review*, vol. 4, no. 5, pp. 193–220, 1890.

[223] D. J. Solove, "The myth of the privacy paradox," *Geo. Wash. L. Rev.*, vol. 89, p. 1, 2022.

[224] N. M. Richards and W. Hartzog, "Taking trust seriously in privacy law," *Stan. Tech. L. Rev.*, vol. 19, p. 431, 2019.

[225] S. Gúrses and J. V. Hoboken, "Privacy after the agile turn," in *The Cambridge handbook of consumer privacy*, pp. 579–601, 2018.

[226] J. Pridmore, "Consumer surveillance, profiling and social sorting: implications for the social fabric," in *The ethics of surveillance*, pp. 201–213, Routledge, 2019.

[227] D. Harwell, "Facial recognition company Clearview AI probed by Canada privacy agencies." Washington Post, 2020.

[228] S. Feldstein, "The global expansion of AI surveillance," 2021.

[229] J. Isaak and M. J. Hanna, "User Data Privacy: Facebook, Cambridge Analytica, and Privacy Protection," *Computer*, vol. 51, no. 8, pp. 56–59, 2018.

[230] E. Gúnel, B. Basaran, and T. Ozyer, "Embedded ethics in surveillance technologies: An analysis of current shortcomings and recommendations," *Philosophy & Technology*, vol. 34, no. 3, pp. 729–769, 2021.

[231] H. Lewis, "Surveillance capitalism and its discontents," *Socialist Lawyer*, no. 89, pp. 8–11, 2021.

[232] M. Stoilova, S. Livingstone, R. Khazbak, and R. Nandagoapa, "Investigating risks of discrimination and exclusion in artificial intelligence for children," 2021.

[233] A. Chander, "The racist algorithm," *Mich. L. Rev.*, vol. 115, p. 1023, 2017.

[234] J. W. Penney, "Chilling effects: Online surveillance and Wikipedia use," *Berkeley Tech. LJ*, vol. 31, p. 117, 2017.

[235] R. Richardson, J. M. Schultz, and K. Crawford, "Dirty data, bad predictions: How civil rights violations impact police data, predictive policing systems, and justice," *New York University Law Review Online, Forthcoming*, 2019.

[236] B. C. Stahl, J. Timmermans, and C. Flick, "Ethics of Emerging Information and Communication Technologies: On the implementation of responsible research and innovation," *Science and Public Policy*, vol. 44, no. 3, pp. 369–381, 2017.

[237] J. Morley, L. Floridi, L. Kinsey, and A. Elhalal, "From What to How: An Initial Review of Publicly Available AI Ethics Tools, Methods and Research to Translate Principles into Practices," *Sci. Eng. Ethics*, vol. 26, no. 4, pp. 2141–2168, 2020.

[238] S. Makridakis, "The forthcoming Artificial Intelligence (AI) revolution: Its impact on society and firms," *Futures*, vol. 90, pp. 46–60, 2017.

[239] R. K. Logan, "Conversational agents and the human workforce of the future," in *Handbook on AI and Human Resource Management*, pp. 1–23, Edward Elgar Publishing, 2022.

[240] E. Brynjolfsson and T. Mitchell, "What can machine learning do? Workforce implications," *Science*, vol. 358, no. 6370, pp. 1530–1534, 2017.

[241] P. R. Daugherty and H. J. Wilson, "Human+Machine: Reimagining Work in the Age of AI," *Harvard Business Review Press*, 2018.

[242] D. Acemoglu and P. Restrepo, "Demographics, Automation and Inequality," 2022.

[243] E. Brynjolfsson, T. Mitchell, and D. Rock, "What Can Machines Learn, and What Does It Mean for Occupations and the Economy?," *AEA Papers and Proceedings*, vol. 112, pp. 43–47, 2022.

[244] D. Autor, "Work of the past, work of the future," *AEA Papers and Proceedings*, vol. 109, pp. 1–32, 2019.

[245] R. Hendra, "Rethinking how Americans transition from high school to college and careers," tech. rep., Community College Research Center Working Paper No. 75, 2013.

[246] A. K. Agrawal, J. S. Gans, and A. Goldfarb, *Prediction Machines: The Simple Economics of Artificial Intelligence*. Harvard Business Press, 2018.

[247] I. M. Cockburn, R. Henderson, and S. Stern, "The impact of artificial intelligence on innovation," Tech. Rep. w24449, National Bureau of Economic Research, 2018.

[248] D. Autor, D. Dorn, L. F. Katz, C. Patterson, and J. Van Reenen, "The fall of the labor share and the rise of superstar firms," *The Quarterly Journal of Economics*, vol. 135, no. 2, pp. 645–709, 2020.

[249] J. Bughin, E. Hazan, S. Ramaswamy, M. Chui, T. Allas, P. Dahlström, N. Henke, and M. Trench, "Artificial intelligence: The next digital frontier?," tech. rep., McKinsey Global Institute, 2017.

[250] D. Rotman, "How technology is destroying jobs," *MIT Technology Review*, vol. 116, no. 4, pp. 28–35, 2013.

[251] D. H. Autor and B. D. Price, "The changing task composition of the US labor market: An update of Autor, Levy, and Murnane (2003)," Working Paper (13-01), MIT Department of Economics, 2013.

[252] K. Bronson and I. Knezevic, "Big Data in food and agriculture," *Big Data & Society*, vol. 3, no. 1, p. 2053951716648174, 2016.

[253] E. Moretti, "The effect of high-tech clusters on the productivity of top inventors," in *AEA Papers and Proceedings*, vol. 109, 2019.

[254] R. Florida, *The new urban crisis: How our cities are increasing inequality, deepening segregation, and failing the middle class–and what we can do about it*. Basic Books, 2017.

[255] D. Rodrik, "Populism and the economics of globalization," *Journal of International Business Policy*, vol. 1, no. 1, pp. 12–33, 2018.

[256] A. Markusen and B. Stucki, "The Trajectory of Regional Economic Development," tech. rep., R/ECON Working Papers, 2020.

[257] A. Caliskan, J. J. Bryson, and A. Narayanan, "Semantics derived automatically from language corpora contain human-like biases," *Science*, vol. 356, no. 6334, pp. 183–186, 2017.

[258] K. Lum and W. Isaac, "To predict and serve?," *Significance*, vol. 13, no. 5, pp. 14–19, 2016.

[259] S. Vaidhyanathan, *Anti-social media: How Facebook disconnects us and undermines democracy*. Oxford University Press, 2018.

[260] V. Eubanks, *Automating Inequality: How High-Tech Tools Profile, Police, and Punish the Poor*. St. Martin's Press, 2018.

[261] P. Restrepo, "Skill and Training Requirements of the Future of Work," in *The Future of Work*, pp. 211–230, Springer, Cham, 2022.

[262] D. Acemoglu and P. Restrepo, "Robots and Jobs: Evidence from US Labor Markets," *Journal of Political Economy*, vol. 128, no. 6, pp. 2188–2244, 2020.

[263] A. Agrawal, A. Goldfarb, and J. S. Gans, eds., *The Economics of Artificial Intelligence: An Agenda*. University of Chicago Press, 2022.

[264] R. Perrault, Y. Shoham, E. Brynjolfsson, J. Clark, J. Etchemendy, B. Grosz, T. Lyons, J. Manyika, S. Mishra, and J. Niebles, "The ai index 2019 annual report," 2019.

[265] OECD, *Getting Skills Right: Future-Ready Adult Learning Systems*. Paris: OECD Publishing, 2019.

[266] P. Restrepo and M. Graff, "Implications of Artificial Intelligence and Digitalization for MENA Labor Markets," *Journal of Economics and Financial Research*, vol. 5, no. 3, pp. 173–190, 2022.

[267] M. Brussevich, C. Dabla-Norris, C. Kamunge, P. Karnane, S. Khalid, and K. Kochhar, "Gender, technology, and the future of work," tech. rep., International Monetary Fund, 2022.

[268] J. M. Arnold, A. Mattoo, and G. Narciso, "Services trade policy and manufacturing productivity: The role of institutions," *Journal of International Economics*, vol. 114, pp. 157–175, 2018.

[269] World Economic Forum, "The future of jobs report 2020," 2020.

[270] A. Rao, G. Verweij, and E. Cameron, "Sizing the prize: What's the real value of AI for your business and how can you capitalise?," tech. rep., PwC, 2017.

[271] PwC, "Workforce of the future: The competing forces shaping 2030," 2018.

[272] A. Korinek and J. E. Stiglitz, "Artificial intelligence and its implications for income distribution and unemployment," in *The economics of artificial intelligence: An agenda*, pp. 349–390, University of Chicago Press, 2021.

[273] L. Floridi, J. Cowls, T. C. King, and M. Taddeo, "How to prevent discriminatory outcomes in machine learning," *Science*, vol. 375, no. 6585, pp. 1099–1101, 2022.

[274] M. Whittaker, K. Crawford, R. Dobbe, G. Fried, E. Kaziunas, V. Mathur, S. West, R. Richardson, J. Schultz, and O. Schwartz, "AI now report 2018," 2018.

[275] "Proposal for a regulation laying down harmonized rules on artificial intelligence," 2021.

[276] "Algorithmic accountability policy toolkit," 2018.

[277] "Building a learning economy: Canada's path towards prosperity and inclusion in the 21st century," 2020.

[278] H. Hoynes and J. Rothstein, "Universal basic income in the US and advanced countries," *Annual Review of Economics*, vol. 11, pp. 929–958, 2019.

[279] J. Manyika, S. Lund, M. Chui, J. Bughin, J. Woetzel, P. Batra, *et al.*, "Jobs lost, jobs gained: Workforce transitions in a time of automation," *McKinsey Global Institute*, 2017.

[280] A. Painter and C. Thoung, "Creative citizen, creative state: the principled and pragmatic case for a Universal Basic Income," *Royal Society of Arts*, 2015.

[281] R. A. Moffitt, "The Deserving Poor, the Family, and the U.S. Welfare System," *Demography*, vol. 52, no. 3, pp. 829–849, 2019.

[282] E. L. Glaeser, "Do transfers lower incentives to acquire knowledge and skills? Evidence from providing unconditional cash for young people," *Labour Economics*, vol. 77, p. 102085, 2022.

[283] I. Marinescu, "No strings attached: The behavioral effects of US unconditional cash transfer programs," Tech. Rep. w24337, National Bureau of Economic Research, 2018.

[284] J. Furman, D. Coyle, A. Fletcher, D. McAuley, and P. Marsden, "Unlocking digital competition. Report of the Digital Competition Expert Panel," tech. rep., HM Treasury, 2019.

[285] L. M. Khan, "Amazon's antitrust paradox," *The Yale Law Journal*, vol. 126, no. 3, pp. 710–805, 2017.

[286] U. Gasser and V. A. F. Almeida, "A Layered Model for AI Governance," *IEEE Internet Comput.*, vol. 21, no. 6, pp. 58–62, 2017.

[287] A. Dafoe, "AI governance: a research agenda," tech. rep., Governance of AI Program, Future of Humanity Institute, University of Oxford, 2018.

[288] H. J. Wilson, P. R. Daugherty, and N. Morini-Bianzino, "The jobs that artificial intelligence will create," *MIT Sloan Management Review*, vol. 58, no. 4, pp. 14–16, 2017.

[289] J. Bughin, "When AI works alongside people," *MIT Sloan Management Review*, 2022.

[290] "Maybe your metrics are useless. here's how to tell.," 2020.

[291] E. Brynjolfsson and A. McAfee, *The Second Machine Age: Work, Progress, and Prosperity in a Time of Brilliant Technologies*. W.W. Norton & Company, 2014.

[292] M. Korn and R. E. Silverman, "Forget b-school, d-school is hot," *Wall Street Journal*, no. 6, 2012.

[293] T. H. Davenport and J. Kirby, "Beyond Automation," *Harvard Business Review*, vol. 93, no. 6, pp. 58–65, 2015.

[294] H. Williams, F. Ceci, and K. Kalyanam, "Artificial intelligence in business: Implications for strategic decision making," *Journal of Business Strategy*, 2021.

[295] H. J. Wilson and P. R. Daugherty, "Collaborative intelligence: humans and AI are joining forces," *Harvard Business Review*, vol. 96, no. 4, pp. 114–123, 2018.

[296] A. S. Grove, *Only the paranoid survive*. Currency, 1999.

[297] J. C. Messenger, O. Vargas Llave, L. Gschwind, S. Boehmer, G. Vermeylen, and M. Wilkens, "Working anytime, anywhere: The effects on the world of work," tech. rep., Eurofound and the International Labour Foundation, 2017.

[298] N. Bloom, "To Raise Productivity, Let More Employees Work from Home," *Harvard Business Review*, 2014.

[299] I. J. Petrick, "Embracing lifelong learning to prepare for the technological disruptions ahead: A policy perspective," *Global Perspectives*, vol. 1, no. 1, 2020.

[300] J. Manyika, M. Chui, J. Bughin, R. Dobbs, P. Bisson, and A. Marrs, *Disruptive technologies: Advances that will transform life, business, and the global economy*, vol. 12. McKinsey Global Institute, 2013.

[301] L. Floridi, J. Cowls, M. Beltrametti, R. Chatila, P. Chazerand, V. Dignum, C. Luetge, R. Madelin, U. Pagallo, F. Rossi, B. Schafer, P. Valcke, and E. Vayena, "AI4People - An Ethical Framework for a Good AI Society: Opportunities, Risks, Principles, and Recommendations," *Minds Mach.*, vol. 28, no. 4, pp. 689–707, 2018.

[302] P. Stone, R. Brooks, E. Brynjolfsson, R. Calo, O. Etzioni, G. Hager, *et al.*, "Artificial intelligence and life in 2030," *One Hundred Year Study on Artificial Intelligence: Report of the 2015-2016 Study Panel*, vol. 52, 2016.

[303] K. Raworth, *Doughnut economics: seven ways to think like a 21st-century economist*. Chelsea Green Publishing, 2018.

[304] F. A. Raso, H. Hilligoss, V. Krishnamurthy, C. Bavitz, and L. Kim, "Artificial Intelligence & Human Rights: Opportunities & Risks," Tech. Rep. 2018-6, Berkman Klein Center Research Publication, 2018.

[305] C. S. Dweck, *Mindset: The new psychology of success*. Random House Digital, Inc., 2008.

[306] O. for Economic Cooperation and D. (OECD), *Skills Matter: Further Results from the Survey of Adult Skills*. OECD Skills Studies, OECD Skills Studies, 2016.

[307] H. Ibarra and M. Hunter, "How leaders create and use networks," *Harvard Business Review*, vol. 85, no. 1, pp. 40–47, 2007.

[308] W. Marcinkus Murphy, "Reverse mentoring at work: Fostering cross-generational learning and developing millennial leaders," *Human Resource Management*, vol. 51, no. 4, pp. 549–573, 2012.

[309] K. Schwab, *The Fourth Industrial Revolution*. Crown Business, 2017.

[310] D. F. Kuratko and R. M. Hodgetts, *Entrepreneurship: Theory, process, practice*. Cengage Learning, 2007.

[311] J. P. Kotter, "What leaders really do," 2021.

[312] J. H. Miller, "Building an understanding of AI ethics," *IT Professional*, vol. 22, no. 6, pp. 58–62, 2020.

[313] J. M. Kouzes and B. Z. Posner, *The leadership challenge*. John Wiley & Sons, 6th ed., 2017.

[314] T. Kochan and L. Dyer, *Shaping the Future of Work: A Handbook for Action and a New Social Contract.* Routledge, 2020.

[315] M. Weijs-Perrée, J. van de Koevering, R. Appel-Meulenbroek, and T. Arentze, "Analysing user preferences for co-working space characteristics," *Building Research & Information*, vol. 47, no. 5, pp. 534–548, 2019.

[316] A. Deutschman, "The internet, open innovation, and the direction of organizational change," *Strategic Management Review*, vol. 1, no. 1, pp. 95–113, 2021.

[317] M. Platt and R. Warwick, "Executives' personalities and AI strategies in innovation: Opening the black box," *Business Horizons*, vol. 62, no. 5, pp. 625–633, 2019.

[318] M. A. Louie, "People and machines: The impact of AI on business leadership," *California Management Review*, vol. 64, no. 1, pp. 5–19, 2021.

[319] T. Fountaine, B. McCarthy, and T. Saleh, "Building the ai-powered organization," *Harvard Business Review*, vol. 97, no. 4.

[320] S. M. West, M. Whittaker, and K. Crawford, "Discriminating Systems: Gender, Race, and Power in AI," *AI Now Institute*, pp. 1–38, 2019.

[321] N. Nilsson, "Artificial intelligence, employment, ethics and policy," *AI and Ethics*, vol. 1, no. 4, pp. 317–326, 2021.

[322] B. Mittelstadt, "Automation, algorithms, and politics ethics of the algorithm: From radical content neutrality to a social constructivist approach," *Philosophy & Technology*, vol. 29, no. 3, pp. 209–219, 2016.

[323] D. Greene, A. L. Hoffmann, and L. Stark, "Better, Nicer, Clearer, Fairer: A Critical Assessment of the Movement for Ethical Artificial Intelligence and Machine Learning," in *52nd Hawaii International Conference on System Sciences, HICSS 2019, Grand Wailea, Maui, Hawaii, USA, January 8-11, 2019* (T. Bui, ed.), pp. 1–10, ScholarSpace, 2019.

[324] E. Shook and K. F. Kee, "Overcoming algorithm aversion: How to develop human-centered algorithms," *Journal of Management Studies*, vol. 59, no. 3, pp. 596–617, 2022.

[325] M. Zeng, "Ethics of artificial intelligence: Connecting people, technology and values," *Current Opinion in Psychology*, vol. 42, pp. 90–95, 2021.

[326] A. Agrawal, J. Gans, and A. Goldfarb, "Exploring the impact of artificial intelligence: Prediction versus judgment," *Information Economics and Policy*, vol. 47, pp. 1–6, 2019.

[327] A. Thierer, W. Crews, C. Sargent, E. Skirmuntt, M. Kotrous, and K. Kovacs, "Comments to national telecommunications and information administration: BroadbandUSA," tech. rep., Mercatus Center, George Mason University, 2017.

[328] S. Popat and L. Starkey, "Learning to code or coding to learn? A systematic review of the evidence on coding in education," *British Journal of Educational Technology*, vol. 50, no. 5, pp. 2166–2182, 2019.

[329] M. Hobday, X. Zhang, and T. Sturgeon, "Mapping regional clusters in China and India: Electronics industry cases," *International Journal of Technological Learning, Innovation and Development*, vol. 5, no. 1-2, pp. 38–65, 2012.

[330] J. Allen, "DARPA and the internet revolution," *Endeavour*, vol. 43, no. 2, pp. 74–82, 2019.

[331] C. Lockhart, S. Austin, C. H. Nally, D. Ramachandran, and M. Walshok, *Emerging best practice for regional talent mobility*. Brookings Institution Press, 2022.

[332] S. G. Benzell, L. J. Kotlikoff, G. LaGarda, and J. D. Sachs, "Robots are us: Some economics of human replacement," tech. rep., NBER working paper 20941, 2015.

[333] G. Colvin, "Taxing robots and artificial intelligence: tax policy solutions to automation risks," *Florida Tax Review*, vol. 25, p. 1, 2021.

[334] J. A. Hauge, M. A. Jamison, and D. Meier, "Subsidizing high-speed internet access for low-income households: Evidence from Minneapolis's pilot program," tech. rep., Bay-Lake Regional Planning Commission, 2020.

[335] A. Tutt, "An FDA for algorithms," *Admin. L. Rev.*, vol. 69, p. 83, 2017.

[336] D. S. Char, N. H. Shah, and D. Magnus, "Implementing Machine Learning in Health Care–Addressing Ethical Challenges," *New England Journal of Medicine*, vol. 378, no. 11, p. 981, 2018.

[337] C. Cath, S. Wachter, B. D. Mittelstadt, M. Taddeo, and L. Floridi, "Artificial Intelligence and the 'Good Society': the US, EU, and UK approach," *Sci. Eng. Ethics*, vol. 24, no. 2, pp. 505–528, 2018.

[338] R. Susskind, *A world without work: Technology, automation and how we should respond*. Metropolitan Books, 2020.

[339] D. Rolnick, P. L. Donti, L. H. Kaack, K. Kochanski, A. Lacoste, K. Sankaran, A. S. Ross, N. Milojevic-Dupont, N. Jaques, A. Waldman-Brown, A. S. Luccioni, T. Maharaj, E. D. Sherwin, S. K. Mukkavilli, K. P.

Kording, C. P. Gomes, A. Y. Ng, D. Hassabis, J. C. Platt, F. Creutzig, J. T. Chayes, and Y. Bengio, "Tackling Climate Change with Machine Learning," *ACM Comput. Surv.*, vol. 55, no. 2, pp. 42:1–42:96, 2023.

[340] R. D. Abbott and B. Bogenschneider, "Should robots pay taxes? Tax policy in the age of automation," *Harvard Law & Policy Review*, vol. 12, p. 145, 2018.

[341] D. Acemoglu and P. Restrepo, "Artificial intelligence, automation and work," no. w24196, 2018.

[342] W. Rosa, N. Gudowsky, and P. Warnke, "But what if? How policymakers use scenario processes to prepare for AI futures," *Futures*, vol. 126, p. 102663, 2021.

[343] L. Kimbell, "Speculative design in government policymaking," *DesignIssues*, vol. 37, no. 2, pp. 64–76, 2021.

[344] M. A. Ali Elfa and M. E.-T. Dawood, "Using Artificial Intelligence for enhancing Human Creativity," *Journal of Art, Design and Music*, vol. 2, no. 2, p. 3, 2023.

[345] Y. N. Harari, *21 lessons for the 21st century*. Random House, 2018.

[346] E. Topol, *Deep medicine: How artificial intelligence can make healthcare human again*. Hachette UK, 2019.

[347] D. Abdullah, R. L. Earley, V. Gold, M. U. Anwar, D. Collings, O. Gencoglu, *et al.*, "Participatory design through a collective intelligence lens: Perspectives and recommendations on designing with marginalized groups," *Human–Computer Interaction*, vol. 37, no. 5-6, pp. 427–463, 2022.

[348] K. Johnson, "The future is individualized learning," *Harvard Business Review*, 2022.

[349] S. Jain and A. Gervais, "The decentralized autonomous organization (DAO): Investor-led governance on the blockchain," *Academy of Management Perspectives*, 2021.

[350] K. Illeris, *Lifelong learning and self: Beyond education for employment*. Routledge, 2021.

[351] J. Diamond, *Guns, Germs, and Steel: The Fates of Human Societies*. W. W. Norton & Company, 1997.

[352] D. S. Landes, *The Unbound Prometheus: Technological Change and Industrial Development in Western Europe from 1750 to the Present*. Cambridge University Press, 1969.

[353] D. Bell, *The Coming of Post-Industrial Society: A Venture in Social Forecasting*. Basic Books, 1973.

[354] A. Kaplan and M. Haenlein, "Siri, Siri, in my hand: Who's the fairest in the land? On the interpretations, illustrations, and implications of artificial intelligence," *Business Horizons*, vol. 62, no. 1, pp. 15–25, 2019.

[355] N. Bostrom, *Superintelligence: Paths, Dangers, Strategies*. Oxford University Press, 2014.

[356] J. Manyika, M. Chui, M. Miremadi, J. Bughin, K. George, P. Willmott, and M. Dewhurst, "A future that works: AI, automation, employment, and productivity," *McKinsey Global Institute Research, Tech. Rep*, vol. 60, pp. 1–135, 2017.

[357] W. Wallach and C. Allen, *Moral Machines: Teaching Robots Right from Wrong*. Oxford University Press, 2009.

[358] C. Homburg, J. Wieseke, and T. Bornemann, "The Impact of Customer Interaction on Customer Satisfaction," *Journal of Marketing*, vol. 79, no. 1, pp. 21–40, 2015.

[359] S. Turkle, *Alone Together: Why We Expect More from Technology and Less from Each Other*. Basic Books, 2011.

[360] R. W. Picard, *Affective Computing*. MIT Press, 1997.

[361] E. Goffman, *The Presentation of Self in Everyday Life*. Doubleday, 1959.

[362] L. Floridi, *The Fourth Revolution: How the Infosphere is Reshaping Human Reality*. Oxford University Press, 2014.

[363] P. M. Napoli, *Social Media and the Public Interest: Media Regulation in the Disinformation Age*. Columbia University Press, 2019.

[364] N. Thurman and S. Schifferes, "The Future of Personalization at News Websites," *Journalism Studies*, vol. 13, no. 5-6, pp. 775–790, 2012.

[365] C. A. Gomez-Uribe and N. Hunt, "The Netflix Recommender System: Algorithms, Business Value, and Innovation," *ACM Trans. Manag. Inf. Syst.*, vol. 6, no. 4, pp. 13:1–13:19, 2016.

[366] J. H. Engel, M. D. Hoffman, and A. Roberts, "Latent Constraints: Learning to Generate Conditionally from Unconditional Generative Models," in *6th International Conference on Learning Representations, ICLR 2018, Vancouver, BC, Canada, April 30 - May 3, 2018, Conference Track Proceedings*, OpenReview.net, 2018.

[367] J. Van Dijck, *The Culture of Connectivity: A Critical History of Social Media*. Oxford University Press, 2013.

[368] E. Wenger, *Communities of Practice: Learning, Meaning, and Identity*. Cambridge University Press, 1998.

[369] J. Bughin, E. Hazan, S. Ramaswamy, M. Chui, T. Allas, P. Dahlstróm, N. Henke, and M. Trench, "The Promise and Challenge of the Age of Artificial Intelligence," *McKinsey Global Institute*, 2018.

[370] V. Mayer-Schónberger and K. Cukier, *Big Data: A Revolution That Will Transform How We Live, Work, and Think*. Houghton Mifflin Harcourt, 2013.

[371] C. Argyris and D. A. Schón, *Organizational Learning: A Theory of Action Perspective*. Addison-Wesley, 1978.

[372] T. H. Davenport and R. Ronanki, "Cognitive Technologies: The Real Opportunities for Business," *Deloitte Review*, 2018.

[373] J. M. Keynes, "Economic Possibilities for our Grandchildren," *Nation and Athenaeum*, vol. 48, pp. 96–102, 1930.

[374] J. Rifkin, *The End of Work: The Decline of the Global Labor Force and the Dawn of the Post-Market Era*. TarcherPerigee, 1995.

[375] J. B. Schor, *The Overworked American: The Unexpected Decline of Leisure*. Basic Books, 1992.

[376] L. Winner, *The Whale and the Reactor: A Search for Limits in an Age of High Technology*. University of Chicago Press, 1986.

[377] D. H. Autor, F. Levy, and R. J. Murnane, "The Skill Content of Recent Technological Change: An Empirical Exploration," *Quarterly Journal of Economics*, vol. 118, no. 4, pp. 1279–1333, 2003.

[378] D. Acemoglu and P. Restrepo, "Automation and New Tasks: How Technology Displaces and Reinstates Labor," *Journal of Economic Perspectives*, vol. 33, no. 2, pp. 3–30, 2018.

[379] R. Bregman, "Utopia for Realists: How We Can Build the Ideal World," *The Correspondent*, 2017.

[380] E. Brynjolfsson, A. McAfee, and M. Spence, "Machines as the Measure of Men: Science, Technology, and Ideologies of Western Dominance," *Foreign Affairs*, vol. 94, 2015.

[381] V. De Stefano, "The Rise of the Just-in-time Workforce: On-demand Work, Crowdwork, and Labor Protection in the Gig Economy," *Comparative Labor Law & Policy Journal*, vol. 37, 2016.

[382] U. Huws, N. H. Spencer, and S. Joyce, "The Future of Work: A Literature Review," *New Technology, Work and Employment*, vol. 34, no. 1, pp. 3–20, 2019.

[383] R. Susskind and D. Susskind, "The Future of the Professions: How Technology Will Transform the Work of Human Experts," *Oxford University Press*, 2015.

[384] K. Schwab, "Shaping the Fourth Industrial Revolution," *World Economic Forum*, 2018.

[385] A. Sundararajan, "From Zipcar to the Sharing Economy," *Harvard Business Review*, 2013.

[386] A. Malhotra and M. W. Van Alstyne, "The Sharing Economy: A Pathway to Sustainability or a Nightmarish Form of Neoliberal Capitalism?," *Communications of the ACM*, vol. 57, no. 4, pp. 32–34, 2014.

[387] G. Zervas, D. Proserpio, and J. W. Byers, "The Rise of the Sharing Economy: Estimating the Impact of Airbnb on the Hotel Industry," *Journal of Marketing Research*, vol. 54, no. 5, pp. 687–705, 2017.

[388] J. V. Hall and A. B. Krueger, "An Analysis of the Labor Market for Uber's Driver-partners in the United States," *ILR Review*, vol. 71, no. 3, pp. 705–732, 2018.

[389] M. Cohen and A. Sundararajan, "Self-regulation and Innovation in the Peer-to-peer Sharing Economy," *University of Chicago Law Review Dialogue*, vol. 82, pp. 116–133, 2015.

[390] D. Guttentag, "Airbnb: Disruptive Innovation and the Rise of an Informal Tourism Accommodation Sector," *Current Issues in Tourism*, vol. 18, no. 12, pp. 1192–1217, 2015.

[391] J. Cramer and A. B. Krueger, "Disruptive Change in the Taxi Business: The Case of Uber," *American Economic Review*, vol. 106, no. 5, pp. 177–182, 2016.

[392] A. Todolí-Signes, "The End of the Subcontracting Model? On-demand Digital Economy Platforms and the Rise of the Crowdworker-prosumer," *Journal of Labor and Society*, vol. 20, no. 4, pp. 479–492, 2017.

[393] C. Blease, L. Fernandez, S. K. Bell, T. Delbanco, and G. Elwyn, "Digital Health and Patient Empowerment: Redefining the Doctor-patient Relationship," *Journal of Medical Internet Research*, vol. 20, no. 3, p. e10631, 2018.

[394] A. Sundararajan, *The Sharing Economy: The End of Employment and the Rise of Crowd-Based Capitalism*. MIT Press, 2016.

[395] J. E. Bessen, "How Computer Automation Affects Occupations: Technology, Jobs, and Skills," *Boston Univ. School of Law, Law and Economics Research Paper*, no. 15-49, 2016.

[396] Z. Obermeyer and E. J. Emanuel, "Predicting the Future – Big Data, Machine Learning, and Clinical Medicine," *New England Journal of Medicine*, vol. 375, no. 13, pp. 1216–1219, 2016.

[397] D. Grewal, A. L. Roggeveen, and J. Nordfält, "Innovations in Retail Pricing and Promotions," *Journal of Retailing*, vol. 93, no. 1, pp. 1–4, 2017.

[398] D. W. Arner, J. N. Barberis, and R. P. Buckley, "FinTech, RegTech, and the Reconceptualization of Financial Regulation," *Northwestern Journal of International Law & Business*, vol. 37, 2016.

[399] K. Liakos, P. Busato, D. Moshou, S. Pearson, and D. Bochtis, "Machine Learning in Agriculture: A Review," *Sensors*, vol. 18, no. 8, p. 2674, 2018.

[400] D. Faggella, "Reskilling and Upskilling for Your Future of Work," *Emerj*, 2018.

[401] M. Broussard, "Artificial Unintelligence: How Computers Misunderstand the World," *MIT Press*, 2018.

[402] K. Crawford and R. Calo, "There is a blind spot in AI research," *Nature*, vol. 538, no. 7625, pp. 311–313, 2016.

[403] J. Bessen, "New Jobs, Old Rules, and AI," *Boston University School of Law, Law and Economics Research Paper*, no. 19-48, 2019.

[404] D. H. Autor, "Why are there still so many jobs? The history and future of workplace automation," *Journal of Economic Perspectives*, vol. 29, no. 3, pp. 3–30, 2015.

[405] J. Bughin, E. Hazan, S. Lund, P. Dahlstróm, A. Wiesinger, and A. Subramaniam, "Skill Shift: Automation and the Future of the Workforce," *McKinsey Global Institute*, 2018.

[406] E. Brynjolfsson and A. McAfee, "Machine, Platform, Crowd: Harnessing Our Digital Future," *WW Norton & Company*, 2017.

[407] D. H. Autor and D. Dorn, "Untangling Trade and Technology: Evidence from Local Labor Markets," *The Economic Journal*, vol. 125, no. 584, pp. 621–646, 2013.

[408] W. E. Forum, "Towards a Reskilling Revolution: A Future of Jobs for All," *World Economic Forum*, 2018.

[409] T. Wagner and T. Dintersmith, *Most Likely to Succeed: Preparing Our Kids for the Innovation Era.* Simon & Schuster, 2015.

[410] R. Luckin, W. Holmes, M. Griffiths, and L. B. Forcier, "Intelligence Unleashed: An Argument for AI in Education," 2016.

[411] W. E. Forum, *The Future of Jobs Report 2018.* World Economic Forum, 2018.

[412] J. E. Bessen, *Learning by Doing: The Real Connection between Innovation, Wages, and Wealth.* Yale University Press, 2017.

[413] A. Agarwal, "Open Learning: The Next Disruptive Innovation," *New England Journal of Higher Education*, 2014.

[414] C. Goldin and L. F. Katz, *The Race between Education and Technology.* Belknap Press, 2018.

[415] "Fosway david perring learning live 2019." `https://www.fosway.com/wp-content/uploads/2019/09/Fosway-Group-Learning-Technology-Landscape-Learning-LIVE-Sept-2019-FINAL.pdf`. (Accessed on 12/27/2023).

[416] "elgpn.eu/publications/browse-by-language/english/elgpn-tools-no-3.-the-evidence-base-on-lifelong-guidance/." `https://www.elgpn.eu/publications/browse-by-language/english/elgpn-tools-no-3.-the-evidence-base-on-lifelong-guidance/`. (Accessed on 12/27/2023).

[417] "Reskilling revolution." `https://initiatives.weforum.org/reskilling-revolution/home`. (Accessed on 12/27/2023).

[418] E. Brynjolfsson, T. Mitchell, and D. Rock, "Work of the Future: Building Better Jobs in an Age of Intelligent Machines," *MIT Task Force on the Work of the Future*, 2020.

[419] J. Reich and J. A. Ruipérez-Valiente, "From MOOCs to Learning at Scale: Research, Methods, and Future Directions," *Annual Review of Statistics and Its Application*, vol. 6, pp. 331–347, 2019.

[420] M. Elfert and K. Rubenson, "Lifelong Learning: Researching a Contested Concept in the Twenty-first Century," in *Third International Handbook of Lifelong Learning*, pp. 1–25, Springer, 2022.

[421] S. Barocas and A. D. Selbst, "Big Data's Disparate Impact," *California Law Review*, vol. 104, p. 671, 2016.

[422] J. Burrell, "How the Machine 'Thinks': Understanding Opacity in Machine Learning Algorithms," *Big Data & Society*, vol. 3, no. 1, p. 2053951715622512, 2016.

[423] A. Cavoukian and J. Jonas, "Privacy by Design: The 7 Foundational Principles," 2012.

[424] M. Taddeo and L. Floridi, "AI and its New Winter: from Myths to Realities," *Philosophy & Technology*, vol. 31, no. 1, pp. 1–3, 2018.

[425] V. Mayer-Schónberger and T. Ramge, *Reinventing Capitalism in the Age of Big Data*. Basic Books, 2018.

[426] K. Martin, "Ethical Implications and Accountability of Algorithms," *Journal of Business Ethics*, vol. 160, no. 4, pp. 835–850, 2019.

[427] P. Voigt and A. Von dem Bussche, *The EU General Data Protection Regulation (GDPR): A Commentary*. Oxford University Press, 2017.

[428] J. A. Van Dijk, *The Digital Divide*. Polity Press, 2017.

[429] L. Robinson, S. R. Cotten, H. Ono, A. Quan-Haase, G. Mesch, W. Chen, J. Schulz, T. M. Hale, and M. J. Stern, "Digital Inequalities and Why They Matter," *Information, Communication & Society*, vol. 18, no. 5, pp. 569–582, 2015.

[430] L. Rainie and J. Anderson, "The future of jobs and jobs training," 2017.

[431] R. D. Atkinson and J. J. Wu, "False alarmism: Technological disruption and the US labor market, 1850–2015," *Information Technology & Innovation Foundation ITIF, May*, 2017.

[432] C. D'Ignazio and L. F. Klein, *Data Feminism*. MIT Press, 2020.

[433] "Ethics guidelines for trustworthy ai shaping europe's digital future." `https://digital-strategy.ec.europa.eu/en/library/ethics-guidelines-trustworthy-ai`. (Accessed on 12/27/2023).

[434] "Inclusive growth, sustainable development and well-being (oecd ai principle) - oecd.ai." `https://oecd.ai/en/dashboards/ai-principles/P5`. (Accessed on 12/27/2023).

[435] "1196_19 beijing consensus_int.indd." `http://www.moe.gov.cn/jyb\`
`_xwfb/gzdt\_gzdt/s5987/201908/W020190828311234688933.`
`pdf`. (Accessed on 12/27/2023).

[436] "About - partnership on ai." `https://partnershiponai.org/about/`.
(Accessed on 12/27/2023).

[437] "World development report 2019: The changing nature of work."
`https://www.worldbank.org/en/publication/wdr2019\#:~:`
`text=The\%20nature\%20of\%20work\%20is\%20changing.`
`\&text=traditional\%20production\%20patterns.-,The\`
`%20rise\%20of\%20the\%20digital\%20platform\%20firm\`
`%20means\%20that\%20technological,\%2Dsolving\%2C\`
`%20teamwork\%20and\%20adaptability`. (Accessed on 12/27/2023).

[438] U. S. Congress, "AI in Government Act of 2019," 2019.

[439] "Artificial intelligence: Development, risks and regulation - house
of lords library." `https://lordslibrary.parliament.uk/`
`artificial-intelligence-development-risks-and-regulation/`.
(Accessed on 12/27/2023).

[440] "Korean-artificial-intelligence-final-report,-innovation,-
brochure." `https://www.intralinkgroup.com/`
`getmedia/7bca58ca-90d0-4c2d-a25f-af23c057b7b3/`
`Korean-Artificial-Intelligence-Final-Report,`
`-Innovation,-Brochure`. (Accessed on 12/27/2023).

[441] IEEE, "Ethically Aligned Design: A Vision for Prioritizing Human Well-
being with Autonomous and Intelligent Systems," 2019.

[442] I. Rahwan, "Society-in-the-Loop: Programming the Algorithmic Social
Contract," *CoRR*, vol. abs/1707.07232, 2017.

[443] M. Tegmark, *Life 3.0: Being Human in the Age of Artificial Intelligence.*
Knopf, 2017.

[444] R. Baldwin, *The Globotics Upheaval: Globalization, Robotics, and the
Future of Work.* Oxford University Press, 2019.

[445] E. Brynjolfsson, D. Rock, and C. Syverson, "Artificial Intelligence and
the Modern Productivity Paradox: A Clash of Expectations and Statistics,"
NBER Working Paper Series, 2018.

[446] U. Nations, "AI for Good Global Summit," 2019.

[447] G. P. on AI, "Global Partnership on Artificial Intelligence," 2020.

[448] R. Kurzweil, *The Singularity is Near: When Humans Transcend Biology*. Penguin, 2016.

[449] Z. M. Vision, "Zebra Medical Vision: AI in Radiology," 2018.

[450] A. Inc., "Atomwise: AI for Drug Discovery," 2019.

[451] IBM, "IBM Watson Health: Pioneering a New Era of Medicine," 2017.

[452] BlueDot, "How AI Predicted the Global Spread of COVID-19," 2020.

[453] S. AG, "Siemens: AI in Manufacturing," 2019.

[454] G. Electric, "Predix: Industrial Internet Platform," 2018.

[455] IBM, "IBM Visual Inspection," 2019.

[456] Amazon, "Amazon Supply Chain Optimization," 2019.

[457] U. Robots, "Universal Robots: Collaborative Industrial Robotic Arms," 2018.

[458] Microsoft, "Microsoft AI Chatbots," 2019.

[459] Amazon, "Amazon Recommendation Engine," 2020.

[460] M. International, "Marriott International: AI in Hospitality," 2019.

[461] J. C. . Co., "JPMorgan Chase: AI in Financial Services," 2019.

[462] U. T. Inc., "Uber: AI in Ride-sharing," 2020.

[463] K. Roose, "This A.I. Subculture's Motto: Go, Go, Go," *The New York Times*, Dec 2023. Accessed: 2023-12-12.

[464] 1kg, "'e/acc': What is Effective Accelerationism?," *Medium*, p. 69, November 2023. Access date: 2023-04-08.

[465] A. Unknown, "Tech Leaders Are Obsessing Over the Obscure Theory E/acc. Here's Why," *Business Insider*, Jul 2023. Accessed: 2023-12-12.

[466] A. Unknown, "Meet the Silicon Valley CEOs Who Say Greed Is Good–Even if It Kills Us All," *Mother Jones*, Dec 2023. Accessed: 2023-12-12.

[467] S. Sellar and D. R. Cole, "Accelerationism: a timely provocation for the critical sociology of education," *British Journal of Sociology of Education*, vol. 38, no. 1, pp. 38–48, 2017.

[468] N. Nilsson, *Principles of Artificial Intelligence*. Springer, 1980.

[469] Y. Hnatchuk, Y. Sierhieiev, and A. Hnatchuk, "Using artificial intelligence accelerators to train computer game characters," *Computer systems and information technologies*, 2021.

[470] D. Acemoglu and P. Restrepo, "The wrong kind of AI? Artificial intelligence and the future of labour demand," *Cambridge Journal of Regions, Economy and Society*, 2019.

[471] N. Corrêa, "The Automation of Acceleration: AI and the Future of Society," *ArXiv*, vol. abs/2007.04477, 2020.

[472] B. Noys, "Days of phuture past: accelerationism in the present moment," *Logos*, vol. 28, pp. 125–136, 2018.

[473] M. E. Gardiner, "Critique of Accelerationism," *Theory, Culture & Society*, vol. 34, pp. 29–52, 2017.

[474] A. T. Williams, "Strategy without a strategiser," *Angelaki*, vol. 24, pp. 14–25, 2019.

[475] P. Haynes, "Is there a future for accelerationism?," *Journal of Organizational Change Management*, 2021.

[476] R. Sheldon, "Accelerationism's queer occulture," *Angelaki*, vol. 24, pp. 118–129, 2019.

[477] B. Rosamond, "Review of Inventing the Future: Postcapitalism and a World without Work by Nick Srnicek and Alex Williams," *New Zealand sociology*, vol. 31, p. 134, 2016.

[478] S. Capozziello, "Curvature Quintessence," 2002.

[479] S. Sellar, "Acceleration, Automation and Pedagogy: How the Prospect of Technological Unemployment Creates New Conditions for Educational Thought," *Education and Technological Unemployment*, 2019.

[480] V. Sahni and L. Wang, "New cosmological model of quintessence and dark matter," *Physical Review D*, vol. 62, p. 103517, 1999.

[481] O. Remaud, "Pequeña filosofía de la aceleración de la historia," *Isegoria*, 2007.

[482] T. Luke, "The Dark Enlightenment and the Anthropocene: Readings from the Book of Third Nature as Political Theology," *Telos*, vol. 2021, pp. 45–68, 2021.

[483] M. Dickerson, "Performative Revolution in Popular Media and Accelerationist Narratives in Resident Evil," *Film Matters*, 2022.

[484] D. Chistyakov, "Philosophy of Accelerationism: A New Way of Comprehending the Present Social Reality (in Nick Land's Context)," *RUDN Journal of Philosophy*, pp. 687–696, 2022.

[485] H. Liu, "On the Accelerationist Critique of the Problem of Capitalist Modernity," *Academic Journal of Humanities & Social Sciences*, 2022.

[486] B. Bernanke, M. Gertler, and S. Gilchrist, "The Financial Accelerator in a Quantitative Business Cycle Framework," *NBER Working Paper Series*, 1998.

[487] G. Sharzer, "Accelerationism and the Limits of Technological Determinism," *Filozofski Vestnik*, vol. 39, 2018.

[488] I. Kapitonov, G. Taspenova, V. Meshkov, and A. Shulus, "Integration of Small and Middle-sized Enterprises into Large Energy Corporations as a Factor of Business Sustainability," *International Journal of Energy Economics and Policy*, vol. 7, pp. 44–52, 2017.

[489] J. Lindl and W. Mead, "Two-dimensional simulation of fluid instability in laser-fusion pellets," *Physical Review Letters*, vol. 34, pp. 1273–1276, 1975.

[490] S. Steenbergen-Hu and S. Moon, "The Effects of Acceleration on High-ability Learners: A Meta-analysis," *Gifted Child Quarterly*, vol. 55, pp. 39–53, 2011.

[491] J. Bin *et al.*, "A laser-driven nanosecond proton source for radiobiological studies," *Applied Physics Letters*, vol. 101, p. 243701, 2012.

[492] L. Suarez-Villa, "The Rise of Technocapitalism," *Science and technology studies*, vol. 28, 2001.

[493] R. Hall, "Platform Discontent against the University," 2020.

[494] R. Kozinets, R. Gambetti, and S. Biraghi, "Faster Than Fact: Consuming in Post-truth Society," *ACR North American Advances*, pp. 413–420, 2018.

[495] A. J. Valentine, "Comment on the multi?person interview: "The future of research universities"," *EMBO reports*, 2007.

[496] R. Kozinets, A. Patterson, and R. Ashman, "Networks of Desire: How Technology Increases Our Passion to Consume," *Journal of Consumer Research*, vol. 43, pp. 659–682, 2016.

[497] P. Callero, "The Globalization of Self: Role and Identity Transformation from Above and Below," *Sociology Compass*, vol. 2, pp. 1972–1988, 2008.

[498] A. Steiber, "Management Characteristics of Top Innovators in Silicon Valley," pp. 23–43, 2018.

[499] A. Berger and A. Brem, "Innovation Hub How-To: Lessons From Silicon Valley," *Global Business and Organizational Excellence*, vol. 35, pp. 58–70, 2016.

[500] S. B. Mahmoud-Jouini, C. Duvert, and M. Esquirol, "Key Factors in Building a Corporate Accelerator Capability," *Research-Technology Management*, vol. 61, pp. 26–34, 2018.

[501] H. Bahrami and S. Evans, "Flexible Re-cycling and High-technology Entrepreneurship," *California Management Review*, vol. 37, pp. 62–89, 1995.

[502] P. Ester, "Accelerators in Silicon Valley," 2018.

[503] J. Haines, "Accelerating cultural capital: reproducing silicon valley culture in global ecosystems," pp. 3–3, 2014.

[504] E. Sutherland and Y. Morieux, "Effectiveness and Competition - Linking Business Strategy, Organizational Culture and the Use of Information Technology," *J. Inf. Technol.*, vol. 3, no. 1, pp. 43–47, 1988.

[505] F. Langerak and E. Hultink, "The impact of new product development acceleration approaches on speed and profitability: lessons for pioneers and fast followers," *IEEE Transactions on Engineering Management*, vol. 52, pp. 30–42, 2005.

[506] M. Carneiro, T. Z. Fulani, and E. Costa, "Práticas e mecanismos de compartilhamento de conhecimento em um programa de aceleração de startups," *Navus: Revista de Gestão e Tecnologia*, vol. 7, pp. 113–123, 2017.

[507] T. Kohler, "Corporate accelerators: Building bridges between corporations and startups," *Business Horizons*, vol. 59, pp. 347–357, 2016.

[508] E. Carayannis and R. Roy, "Davids vs Goliaths in the small satellite industry: - the role of technological innovation dynamics in firm competitiveness," *Technovation*, vol. 20, pp. 287–297, 2000.

[509] J. Barker, "Slow Down," *Angelaki*, vol. 21, pp. 227–235, 2016.

[510] N. Karagozoglu and W. B. Brown, "Time Based Management of the New Product Development Process," *Journal of Product Innovation Management*, vol. 10, pp. 204–215, 1993.

[511] R. Tutton, "Sociotechnical Imaginaries and Techno-Optimism: Examining Outer Space Utopias of Silicon Valley," *Science as Culture*, vol. 30, pp. 416–439, 2020.

[512] N. Land, *ACCELERATE: The Accelerationist Reader*. Urbanomic, 2014.

[513] R. Mackay and A. Avanessian, eds., *ACCELERATE: The Accelerationist Reader*. Urbanomic, 2014.

[514] N. Srnicek and A. Williams, *Inventing the Future: Postcapitalism and a World Without Work*. Verso, 2015.

[515] S. Shaviro, "Accelerationist Aesthetics: Necessary Inefficiency in Times of Real Subsumption," *e-flux journal*, no. 46, 2013.

[516] B. Noys, *Malign Velocities: Accelerationism & Capitalism*. Zero Books, 2014.

[517] R. Keucheyan, *The Left Hemisphere: Mapping Critical Theory Today*. Verso, 2013.

[518] H. Chowdhury, "Get the lowdown on 'e/acc' – Silicon Valley's favorite obscure theory about progress at all costs," 2023.

[519] R. A. Chance, "Meet the Silicon Valley CEOs Who Say Greed Is Good– Even if It Kills Us All "Effective accelerationism" wracked OpenAI. That's just the beginning," *Mother Jones*, December 2023. Online article.

[520] K. Roose, "The Shift. This A.I. Subculture's Motto: Go, Go, Go," *The New York Times*, December 2023.

[521] A. Breland, "The SHIFT This A.I. Subculture's Motto: Go, Go, Go," *Unknown Journal*, December 2023. Online article.

[522] "Effective accelerationism discourse with masad, amjad." Social Media Conversation, 2023.

[523] "Critique of effective accelerationism by park, peter s.." Social Media Commentary, 2023.

[524] "Julie fredrickson on the societal impact of effective accelerationism." Blog Post, 2023.

[525] G. Verdon, "The Role of AI Hardware in Accelerationist Philosophy," *Forbes Tech*, March 2023.

[526] K. Roose, "This A.I. Subculture's Motto: Go, Go, Go," *The New York Times: The Shift*, December 2023. Online; accessed 12 April 2023.

[527] R. A. Chance, "Meet the Silicon Valley CEOs Who Say Greed Is Good– even if It Kills Us All," *Mother Jones*, December 2023. Online; accessed 10 April 2023.

[528] Unnamed Source from AI development community, "The evolution of GPT-4 and the acceleration debate in AI progress," 2023. Personal communication.

[529] J. Fredrickson, "'e/acc': What is Effective Accelerationism?," 2023.

[530] P. S. Park, "On accelerationism – decolonizing technoscience through critical pedagogy," *Journal for Activism in Science & Technology Education*, vol. 6, no. 1, pp. 20–27, 2023.

[531] A. Masad, "'AI Progress and Its Discontents: Perspectives from the Replit CEO'," *TechCrunch*, 2023.

[532] A. Touraine, *Beyond Neoliberalism*. Polity Press, 2001.

[533] B. Noys, "Crash and Burn: Debating Accelerationism," *Online*, 2014. http://www.3ammagazine.com/3am/crash-and-burn-debating-accelerationism/.

[534] F. Beradi, *After the Future*. Oakland, CA: AK Press, 2014.

[535] M. Wark, *Molecular Red: Theory for the Anthropocene*. Verso, 2015.

[536] R. Braidotti, *The Posthuman*. Polity, 2013.

[537] N. Srnicek and A. Williams, "Accelerate: Manifesto for an accelerationist politics," *Critical Legal Thinking*, 2013. http://criticallegalthinking.com/2013/05/14/accelerate-manifesto-for-an-accelerationist-politics/.

[538] A. Masad, "Artificial Intelligence – The Revolution Hasn't Happened Yet," *Harvard Business Review*, 2016. https://hbr.org/2018/07/artificial-intelligence-the-revolution-hasnt-happened-yet.

[539] M. Pasquinelli, "Code Surplus Value and the Augmented Intellect," 2014. http://matteopasquinelli.com/code-surplus-value/.

[540] J. Sullivan and H. Chowdhury, "Understanding Effective Accelerationism: The Manifesto of BasedBeffJezos," *Insider Tech*, pp. 1–10, 2023.

[541] R. Brassier, "Prometheanism and its critics," *Accelerate: The Accelerationist Reader*, pp. 467–489, 2014.

[542] A. Breland, "Meet the Silicon Valley CEOs Who Say Greed Is Good– Even if It Crushes Us All," December 2023.

[543] K. Roose, "The Ethical Accelerationism Dilemma in AI's Future," *The New York Times*, pp. A1–A2, 2023.

[544] R. Mackay and A. Avanessian, *ACCELERATE: The Accelerationist Reader*. Urbanomic, 2014.

[545] B. Noys, "Malign Velocities: Accelerationism and Capitalism," *Zero Books*, 2014. Available at: [URL] Accessed: [Date of access].

[546] N. Srnicek and A. Williams, *Inventing the Future: Postcapitalism and a World Without Work*. Verso Books, 2015.

[547] M. Pasquinelli, "Code Surplus Value and the Augmented Intellect." http://matteopasquinelli.com/code-surplus-value/, 2014.

[548] E. Brynjolfsson and A. McAfee, *The Second Machine Age: Work, Progress, and Prosperity in a Time of Brilliant Technologies*. W. W. Norton & Company, 2014.

[549] S. Russell, *Human Compatible: Artificial Intelligence and the Problem of Control*. Penguin, 2019.

[550] H. Rosa, *Social Acceleration: A New Theory of Modernity*. Columbia University Press, 2013.

[551] D. R. Cole and J. P. Bradley, "Towards a New Critical Ecology: Accelerative Capitalism and a Critical Pedagogy of the Posthuman," *Critical Sociology*, vol. 43, no. 4-5, pp. 621–639, 2017.

[552] A. Breland, "Meet the Silicon Valley CEOs Who Say Greed Is Good–Even if It Kills Us All," *Mother Jones*, 2023.

[553] N. Land, "Templexity: Disordered Loops through Shanghai Time," 2014.

[554] N. Bostrom, *Global Catastrophic Risks*. Oxford University Press, 2016.

[555] S. Zuboff, *The Age of Surveillance Capitalism*. PublicAffairs, 2019.

[556] N. Land, *Fanged Noumena: Collected Writings 1987-2007*. Falmouth, UK: Urbanomic Media Ltd, 2011.

[557] R. A. Chance, "Politics: Meet the Silicon Valley CEOs Who Say Greed Is Good–Even if It Kills Us All," *Mother Jones*, 2023.

[558] R. A. Rosen, "This A.I. Subculture's Motto: Go, Go, Go," 2023.

[559] M. B. Kronenberg, "Accelerationist Aesthetics: Necessary Inefficiency in Times of Real Subsumption," *e-flux journal*, 2014.

[560] R. Shen, "AI Subculture's Motto: Go, Go, Go," *Unpublished manuscript*, 2023.

[561] A. Masad, "Keep A.I. Open," 2023.

[562] M. Andreessen, "The Techno-optimist Manifesto," 2022. Available at: [URL] Accessed: [Date of access].

[563] R. Mackay and A. Avanessian, *ACCELERATE The Accelerationist Reader*. Falmouth, UK: Urbanomic, 2014.

[564] G. Tan, "Effective Accelerationism Panel Discussion," 2023. Available at: [URL] Accessed: [Date of access].

[565] S. N. Visser *et al.*, "Treatment of attention-deficit/hyperactivity disorder among children with special health care needs," *Journal of Pediatrics*, 2015. Published online April 1, 2015.

[566] A. Breland, "Politics Meet the Silicon Valley CEOs Who Say Greed Is Good–even if It Kills Us All "Effective accelerationism" wracked OpenAI. That's just the beginning," *Mother Jones*, vol. 6, no. 12, 2023. Available at: https://www.motherjones.com/politics/2023/12/effective-accelerationism-silicon-valley-tech-ai-andreessen/.

[567] K. Roose, "This A.I. Subculture's Motto: Go, Go, Go The eccentric pro-tech movement known as "Effective Accelerationism" wants to unshackle powerful A.I., and party along the way," *The New York Times*, December 2023. Available at: https://www.nytimes.com/2023/12/10/business/effective-accelerationism-ai.html.

[568] A. Berland, "This A.I. Subculture's Motto: Go, Go, Go," *Mother Jones*, 2023.

[569] M. Andreessen, "A16z VC Firm's Support for AI and e/acc," 2023. Access date: 2023-04-08.

[570] J. Fredrickson, "Silicon Valley Being Down for the Potential Extinction of Humanity Sounds Patently Absurd," *Mother Jones*, 2023.

[571] A. Masad and O. Contributors, "Reflections on e/acc movement and technological optimism," *Social Media Discourse*, 2023. Access date: 2023-04-08.

[572] G. Deleuze and F. Guattari, *Anti-Oedipus: Capitalism and Schizophrenia*. Minneapolis, MN: University of Minnesota Press, 1983.

[573] B. Noys, *Malign Velocities: Accelerationism and Capitalism*. Zero Books, 2014.

[574] F. B. Berardi, *After the future*. AK Press, 2011.

[575] S. J. Russell, *Sapiens rule: The ethics of algorithms and how we control autonomous systems*. Cambridge University Press, 2015.

[576] M. Jones, "This A.I. Subculture's Motto: Go, Go, Go," *Mother Jones*, December 2023.

[577] R. Negarestani, "The Labour of the Inhuman," *Urbanomic*, 2014.

[578] P. Wolfendale, "Prometheanism and rationalism," *Academia*, 2016.

[579] E. S. Marques, C. L. d. Moraes, M. H. Hasselmann, S. F. Deslandes, and M. E. Reichenheim, "Violence against women, children, and adolescents during the COVID-19 pandemic: overview, contributing factors, and mitigating measures," *Cadernos de saude publica*, vol. 36, 2020.

[580] S. Maital and E. Barzani, "The global economic impact of COVID-19: A summary of research," *Samuel Neaman Institute for National Policy Research*, vol. 2020, pp. 1–12, 2020.

[581] K. S. Khan, M. A. Mamun, M. D. Griffiths, and I. Ullah, "The mental health impact of the COVID-19 pandemic across different cohorts," *International journal of mental health and addiction*, vol. 20, no. 1, pp. 380–386, 2022.

[582] W. Msemburi, A. Karlinsky, V. Knutson, S. Aleshin-Guendel, S. Chatterji, and J. Wakefield, "The WHO estimates of excess mortality associated with the COVID-19 pandemic," *Nature*, vol. 613, no. 7942, pp. 130–137, 2023.

[583] Y. Assefa, C. F. Gilks, R. van de Pas, S. Reid, D. G. Gete, and W. Van Damme, "Reimagining global health systems for the 21st century: lessons from the COVID-19 pandemic," *BMJ global health*, vol. 6, no. 4, p. e004882, 2021.

[584] R. Vaishya, M. Javaid, I. H. Khan, and A. Haleem, "Artificial Intelligence (AI) applications for COVID-19 pandemic," *Diabetes & Metabolic Syndrome: Clinical Research & Reviews*, vol. 14, no. 4, pp. 337–339, 2020.

[585] N. Darapaneni, A. Sreevanth, A. R. Paduri, R. Sivakumaran, S. Sundaramurthi, *et al.*, "Explainable diagnosis, lesion segmentation and quantification of COVID-19 infection from CT images using convolutional neural networks," in *2022 IEEE 13th Annual Information Technology, Electronics and Mobile Communication Conference (IEMCON)*, pp. 0171–0178, IEEE, 2022.

[586] S. Huang, J. Yang, S. Fong, and Q. Zhao, "Artificial intelligence in the diagnosis of COVID-19: challenges and perspectives," *International journal of biological sciences*, vol. 17, no. 6, p. 1581, 2021.

[587] C. Silva and D. Saraee, "Literature review on epidemiological modelling, spatial modelling and artificial intelligence for COVID-19," *Journal of Advances in Medicine and Medical Research*, vol. 33, no. 5, pp. 8–21.

[588] R. Vicari and N. Komendatova, "Systematic meta-analysis of research on AI tools to deal with misinformation on social media during natural and anthropogenic hazards and disasters," *Humanities and Social Sciences Communications*, vol. 10, no. 1, pp. 1–14, 2023.

[589] E. Al Sulais, M. Mosli, and T. AlAmeel, "The psychological impact of COVID-19 pandemic on physicians in Saudi Arabia: a cross-sectional study," *Saudi journal of gastroenterology: official journal of the Saudi Gastroenterology Association*, vol. 26, no. 5, p. 249, 2020.

[590] Y.-S. R. Poon, Y. P. Lin, P. Griffiths, K. K. Yong, B. Seah, and S. Y. Liaw, "A global overview of healthcare workers' turnover intention amid COVID-19 pandemic: A systematic review with future directions," *Human resources for health*, vol. 20, no. 1, pp. 1–18, 2022.

[591] D. Mhlanga, "The role of artificial intelligence and machine learning amid the COVID-19 pandemic: What lessons are we learning on 4IR and the sustainable development goals," *International Journal of Environmental Research and Public Health*, vol. 19, no. 3, p. 1879, 2022.

[592] M. K. Yogi and J. Garikipati, "Future Scope of Artificial Intelligence in Healthcare for COVID-19," *Emerging Technologies for Combatting Pandemics*, pp. 85–100, 2022.

[593] Q. Pham, D. C. Nguyen, T. Huynh-The, W. Hwang, and P. N. Pathirana, "Artificial Intelligence (AI) and Big Data for Coronavirus (COVID-19) Pandemic: A Survey on the State-of-the-arts," *CoRR*, vol. abs/2107.14040, 2021.

[594] W. Naudé, "Artificial Intelligence against COVID-19: An early review," 2020.

[595] A. S. Adly, A. S. Adly, and M. S. Adly, "Approaches based on artificial intelligence and the internet of intelligent things to prevent the spread of COVID-19: scoping review," *Journal of medical Internet research*, vol. 22, no. 8, p. e19104, 2020.

[596] M. J. Page, J. E. McKenzie, P. M. Bossuyt, I. Boutron, T. C. Hoffmann, C. D. Mulrow, L. Shamseer, J. M. Tetzlaff, E. A. Akl, S. E. Brennan, *et al.*, "The PRISMA 2020 statement: an updated guideline for reporting systematic reviews," *International journal of surgery*, vol. 88, p. 105906, 2021.

[597] Y. Li, L. Cao, Z. Zhang, L. Hou, Y. Qin, X. Hui, J. Li, H. Zhao, G. Cui, X. Cui, *et al.*, "Reporting and methodological quality of COVID-19 systematic reviews needs to be improved: an evidence mapping," *Journal of clinical epidemiology*, vol. 135, pp. 17–28, 2021.

[598] F. Jiang, Y. Jiang, H. Zhi, Y. Dong, H. Li, S. Ma, Y. Wang, Q. Dong, H. Shen, and Y. Wang, "Artificial intelligence in healthcare: past, present and future," *Stroke and vascular neurology*, vol. 2, no. 4, 2017.

[599] P. Hamet and J. Tremblay, "Artificial intelligence in medicine," *Metabolism*, vol. 69, pp. S36–S40, 2017.

[600] E. H. Shortliffe, *MYCIN: A rule-based computer program for advising physicians regarding antimicrobial therapy selection.* PhD thesis, Stanford University Ph. D. dissertation, 1974.

[601] E. H. Shortliffe, R. Davis, S. G. Axline, B. G. Buchanan, C. C. Green, and S. N. Cohen, "Computer-based consultations in clinical therapeutics: explanation and rule acquisition capabilities of the MYCIN system," *Computers and biomedical research*, vol. 8, no. 4, pp. 303–320, 1975.

[602] R. A. Miller, H. E. Pople Jr, and J. D. Myers, "Internist-I, an experimental computer-based diagnostic consultant for general internal medicine," in *Computer-assisted medical decision making*, pp. 139–158, Springer, 1985.

[603] A. P. Bradley, *Machine learning for medical diagnostics: Techniques for feature extraction, classification, and evaluation.* The University of Queensland (Australia), 1996.

[604] I. Kononenko, "Inductive and Bayesian learning in medical diagnosis," *Appl. Artif. Intell.*, vol. 7, no. 4, pp. 317–337, 1993.

[605] B. Zupan, J. Demsar, M. W. Kattan, J. R. Beck, and I. Bratko, "Machine learning for survival analysis: a case study on recurrence of prostate cancer," *Artif. Intell. Medicine*, vol. 20, no. 1, pp. 59–75, 2000.

[606] A. Miller, B. Blott, and T. Hames, "Review of neural network applications in medical imaging and signal processing," *Medical and Biological Engineering and Computing*, vol. 30, pp. 449–464, 1992.

[607] S. B. Lo, H. Chan, J. Lin, H. Li, M. T. Freedman, and S. K. Mun, "Artificial convolution neural network for medical image pattern recognition," *Neural Networks*, vol. 8, no. 7-8, pp. 1201–1214, 1995.

[608] P. Larranaga, B. Calvo, R. Santana, C. Bielza, J. Galdiano, I. Inza, J. A. Lozano, R. Armananzas, G. Santafé, A. Pérez, *et al.*, "Machine learning in bioinformatics," *Briefings in bioinformatics*, vol. 7, no. 1, pp. 86–112, 2006.

[609] W. Dubitzky, M. Granzow, and D. P. Berrar, *Fundamentals of data mining in genomics and proteomics.* Springer Science & Business Media, 2007.

[610] W. S. Hayes and M. Borodovsky, "How to interpret an anonymous bacterial genome: machine learning approach to gene identification," *Genome research*, vol. 8, no. 11, pp. 1154–1171, 1998.

[611] A. Zhavoronkov, Y. A. Ivanenkov, A. Aliper, M. S. Veselov, V. A. Aladinskiy, A. V. Aladinskaya, V. A. Terentiev, D. A. Polykovskiy, M. D. Kuznetsov, A. Asadulaev, *et al.*, "Deep learning enables rapid identification of potent DDR1 kinase inhibitors," *Nature Biotechnology*, vol. 37, no. 9, pp. 1038–1040, 2019.

[612] R. Burbidge, M. W. B. Trotter, B. F. Buxton, and S. B. Holden, "Drug Design by Machine Learning: Support Vector Machines for Pharmaceutical Data Analysis," *Comput. Chem.*, vol. 26, no. 1, pp. 5–14, 2002.

[613] V. V. Zernov, K. V. Balakin, A. A. Ivaschenko, N. P. Savchuk, and I. V. Pletnev, "Drug discovery using support vector machines. The case studies of drug-likeness, agrochemical-likeness, and enzyme inhibition predictions," *Journal of chemical information and computer sciences*, vol. 43, no. 6, pp. 2048–2056, 2003.

[614] F. S. Collins, M. Morgan, and A. Patrinos, "The Human Genome Project: lessons from large-scale biology," *Science*, vol. 300, no. 5617, pp. 286–290, 2003.

[615] I. Wallach, M. Dzamba, and A. Heifets, "AtomNet: A Deep Convolutional Neural Network for Bioactivity Prediction in Structure-based Drug Discovery," *CoRR*, vol. abs/1510.02855, 2015.

[616] L. Piwek, D. A. Ellis, S. Andrews, and A. Joinson, "The rise of consumer health wearables: promises and barriers," *PLOS Medicine*, vol. 13, no. 2, p. e1001953, 2016.

[617] G. Kaur, "Pandemic Management via Technology: A Review," *Management*, vol. 40, no. 2, pp. 181–187, 2011.

[618] A. Sadilek, H. A. Kautz, and V. Silenzio, "Predicting Disease Transmission from Geo-tagged Micro-blog Data," in *Proceedings of the Twenty-Sixth AAAI Conference on Artificial Intelligence, July 22-26, 2012, Toronto, Ontario, Canada* (J. Hoffmann and B. Selman, eds.), pp. 136–142, AAAI Press, 2012.

[619] S. S. R. Abidi and A. Goh, "Applying Knowledge Discovery to Predict Infectious Disease Epidemics," in *PRICAI'98, Topics in Artificial Intelligence, 5th Pacific Rim International Conference on Artificial Intelligence, Singapore, November 22-27, 1998, Proceedings* (H. Lee and H. Motoda, eds.), vol. 1531 of *Lecture Notes in Computer Science*, pp. 170–181, Springer, 1998.

[620] V. Lampos and N. Cristianini, "Tracking the flu pandemic by monitoring the social web," in *2nd International Workshop on Cognitive Information Processing, CIP 2010, Elba, Italy, 14-16 June, 2010*, pp. 411–416, IEEE, 2010.

[621] S. Choi, J. Lee, M.-G. Kang, H. Min, Y.-S. Chang, and S. Yoon, "Large-scale machine learning of media outlets for understanding public reactions to nation-wide viral infection outbreaks," *Methods*, vol. 129, pp. 50–59, 2017.

[622] N. L. Bragazzi, H. Dai, G. Damiani, M. Behzadifar, M. Martini, and J. Wu, "How big data and artificial intelligence can help better manage the COVID-19 pandemic," *International journal of environmental research and public health*, vol. 17, no. 9, p. 3176, 2020.

[623] D. Dong, Z. Tang, S. Wang, H. Hui, L. Gong, Y. Lu, Z. Xue, H. Liao, F. Chen, F. Yang, *et al.*, "The role of imaging in the detection and management of COVID-19: a review," *IEEE reviews in biomedical engineering*, vol. 14, pp. 16–29, 2020.

[624] L. Wang, Z. Q. Lin, and A. Wong, "Covid-net: A tailored deep convolutional neural network design for detection of covid-19 cases from chest x-ray images," *Scientific reports*, vol. 10, no. 1, p. 19549, 2020.

[625] T. Ozturk, M. Talo, E. A. Yildirim, U. B. Baloglu, Ö. Yildirim, and U. R. Acharya, "Automated detection of COVID-19 cases using deep neural networks with X-ray images," *Comput. Biol. Medicine*, vol. 121, p. 103792, 2020.

[626] L. Li, L. Qin, Z. Xu, Y. Yin, X. Wang, B. Kong, J. Bai, Y. Lu, Z. Fang, Q. Song, K. Cao, D. Liu, G. Wang, Q. Xu, X. Fang, S. Zhang, J. Xia, and J. Xia, "Artificial intelligence distinguishes COVID-19 from community acquired pneumonia on chest CT," *Radiology*, vol. 296, no. 2, pp. E65–E71, 2020.

[627] F. Shi, L. Xia, F. Shan, B. Song, D. Wu, Y. Wei, H. Yuan, H. Jiang, Y. He, Y. Gao, *et al.*, "Large-scale screening to distinguish between COVID-19 and community-acquired pneumonia using infection size-aware classification," *Physics in medicine & Biology*, vol. 66, no. 6, p. 065031, 2021.

[628] K. Zhang, X. Liu, J. Shen, Z. Li, Y. Sang, X. Wu, Y. Zha, W. Liang, C. Wang, K. Wang, *et al.*, "Clinically applicable AI system for accurate diagnosis, quantitative measurements, and prognosis of COVID-19 pneumonia using computed tomography," *Cell*, vol. 181, no. 6, pp. 1423–1433, 2020.

[629] X. Jiang, M. Coffee, A. Bari, J. Wang, X. Jiang, J. Huang, J. Shi, J. Dai, J. Cai, T. Zhang, *et al.*, "Towards an artificial intelligence framework for data-driven prediction of coronavirus clinical severity," *Computers, Materials & Continua*, vol. 63, no. 1, pp. 537–551, 2020.

[630] A. S. Miner, L. Laranjo, and A. B. Kocaballi, "Chatbots in the fight against the COVID-19 pandemic," *npj Digit. Medicine*, vol. 3, 2020.

[631] M. Almalki and F. Azeez, "Health chatbots for fighting COVID-19: a scoping review," *Acta Informatica Medica*, vol. 28, no. 4, p. 241, 2020.

[632] A. Martin, J. Nateqi, S. Gruarin, N. Munsch, I. Abdarahmane, M. Zobel, and B. Knapp, "An artificial intelligence-based first-line defence against COVID-19: digitally screening citizens for risks via a chatbot," *Scientific reports*, vol. 10, no. 1, p. 19012, 2020.

[633] E. M. Boucher, N. R. Harake, H. E. Ward, S. E. Stoeckl, J. Vargas, J. Minkel, A. C. Parks, and R. Zilca, "Artificially intelligent chatbots in digital mental health interventions: a review," *Expert Review of Medical Devices*, vol. 18, no. sup1, pp. 37–49, 2021.

[634] A. Channa, N. Popescu, J. Skibinska, and R. Burget, "The Rise of Wearable Devices during the COVID-19 Pandemic: A Systematic Review," *Sensors*, vol. 21, no. 17, p. 5787, 2021.

[635] T. Mishra, M. Wang, A. A. Metwally, G. K. Bogu, A. W. Brooks, A. Bahmani, A. Alavi, A. Celli, A. de la Zerda, D. Salins, and M. P. Snyder, "Pre-symptomatic detection of COVID-19 from smartwatch data," *Nature Biomedical Engineering*, vol. 4, no. 12, pp. 1208–1220, 2020.

[636] K. Gandla, K. T. K. Reddy, P. V. Babu, R. Sagapola, and P. Sudhakar, "A REVIEW OF ARTIFICIAL INTELLIGENCE IN TREATMENT OF COVID-19," *Journal of Pharmaceutical Negative Results*, pp. 254–264, 2022.

[637] M. D. Smith and J. C. Smith, "Repurposing Therapeutics for COVID-19: Supercomputer-based Docking to the SARS-CoV-2 Viral Spike Protein and Viral Spike Protein-human ACE2 Interface," *ChemRxiv*, 2020.

[638] Y. Zhou, F. Wang, J. Tang, R. Nussinov, and F. Cheng, "Artificial intelligence in COVID-19 drug repurposing," *The Lancet Digital Health*, vol. 2, no. 12, pp. e667–e676, 2020.

[639] A. W. Senior, R. Evans, J. Jumper, J. Kirkpatrick, L. Sifre, T. Green, C. Qin, A. Zídek, A. W. R. Nelson, A. Bridgland, H. Penedones, S. Petersen, K. Simonyan, S. Crossan, P. Kohli, D. T. Jones, D. Silver, K. Kavukcuoglu, and D. Hassabis, "Improved protein structure prediction

using potentials from deep learning," *Nat.*, vol. 577, no. 7792, pp. 706–710, 2020.

[640] W. D. Jang, S. Jeon, S. Kim, and S. Y. Lee, "Drugs repurposed for COVID-19 by virtual screening of 6,218 drugs and cell-based assay," *Proceedings of the National Academy of Sciences*, vol. 118, no. 30, p. e2024302118, 2021.

[641] M. Kandeel and M. Al-Nazawi, "Virtual screening and repurposing of FDA approved drugs against COVID-19 main protease," *Life sciences*, vol. 251, p. 117627, 2020.

[642] H. Rohmetra, N. Raghunath, P. Narang, V. Chamola, M. Guizani, and N. R. Lakkaniga, "AI-enabled remote monitoring of vital signs for COVID-19: methods, prospects and challenges," *Computing*, vol. 105, no. 4, pp. 783–809, 2023.

[643] A. J. Bokolo, "Application of telemedicine and eHealth technology for clinical services in response to COVID-19 pandemic," *Health and technology*, vol. 11, no. 2, pp. 359–366, 2021.

[644] P. Webster, "Virtual health care in the era of COVID-19," *The Lancet*, vol. 395, no. 10231, pp. 1180–1181, 2020.

[645] T. J. Judson, A. Y. Odisho, J. J. Young, O. Bigazzi, D. Steuer, R. Gonzales, and A. B. Neinstein, "Implementation of a digital chatbot to screen health system employees during the COVID-19 pandemic," *J. Am. Medical Informatics Assoc.*, vol. 27, no. 9, pp. 1450–1455, 2020.

[646] M. Chadi and H. Mousannif, "A Reinforcement Learning Based Decision Support Tool for Epidemic Control: Validation Study for COVID-19," *Appl. Artif. Intell.*, vol. 36, no. 1, 2022.

[647] X. Guo, P. Chen, S. Liang, Z. Jiao, L. Li, J. Yan, Y. Huang, Y. Liu, and W. Fan, "PaCAR: COVID-19 pandemic control decision making via large-scale agent-based modeling and deep reinforcement learning," *Medical Decision Making*, vol. 42, no. 8, pp. 1064–1077, 2022.

[648] M. Ghahramani and F. Pilla, "Leveraging Artificial Intelligence to Analyze the COVID-19 Distribution Pattern based on Socio-economic Determinants," *CoRR*, vol. abs/2102.06656, 2021.

[649] K. Narayan, H. Rathore, and F. Znidi, "Using Epidemic Modeling, Machine Learning and Control Feedback Strategy for Policy Management of COVID-19," *IEEE Access*, vol. 10, pp. 98244–98258, 2022.

[650] H. Wang, "The Application of Artificial Intelligence in Health Care Resource Allocation Before and During the COVID-19 Pandemic: Scoping," *health policy*, vol. 5, p. 6, 2023.

[651] L. Yu, A. Halalau, B. Dalal, A. E. Abbas, F. Ivascu, M. Amin, and G. B. Nair, "Machine learning methods to predict mechanical ventilation and mortality in patients with COVID-19," *PLoS One*, vol. 16, no. 4, p. e0249285, 2021.

[652] N. J. Douville, C. B. Douville, G. Mentz, M. R. Mathis, C. Pancaro, K. K. Tremper, and M. Engoren, "Clinically applicable approach for predicting mechanical ventilation in patients with COVID-19," *British journal of anaesthesia*, vol. 126, no. 3, pp. 578–589, 2021.

[653] M. Cinelli, W. Quattrociocchi, A. Galeazzi, C. M. Valensise, E. Brugnoli, A. L. Schmidt, P. Zola, F. Zollo, and A. Scala, "The COVID-19 Social Media Infodemic," *CoRR*, vol. abs/2003.05004, 2020.

[654] C. M. Pulido, B. Villarejo-Carballido, G. Redondo-Sama, and A. Gómez, "COVID-19 infodemic: More retweets for science-based information on coronavirus than for false information," *International sociology*, vol. 35, no. 4, pp. 377–392, 2020.

[655] A. Z. Klein, A. Magge, K. O'Connor, J. I. Flores Amaro, D. Weissenbacher, and G. Gonzalez Hernandez, "Toward using Twitter for tracking COVID-19: a natural language processing pipeline and exploratory data set," *Journal of medical Internet research*, vol. 23, no. 1, p. e25314, 2021.

[656] G. Newlands, C. Lutz, A. Tamò-Larrieux, E. Fosch-Villaronga, R. Harasgama, and G. Scheitlin, "Innovation under pressure: Implications for data privacy during the Covid-19 pandemic," *Big Data Soc.*, vol. 7, no. 2, p. 205395172097668, 2020.

[657] M. Christofidou, N. Lea, and P. Coorevits, "A literature review on the GDPR, COVID-19 and the ethical considerations of data protection during a time of crisis," *Yearbook of medical informatics*, vol. 30, no. 01, pp. 226–232, 2021.

[658] J. Delgado, A. de Manuel, I. Parra, C. Moyano, J. Rueda, A. Guersenzvaig, T. Ausin, M. Cruz, D. Casacuberta, and A. Puyol, "Bias in algorithms of AI systems developed for COVID-19: A scoping review," *Journal of Bioethical Inquiry*, vol. 19, no. 3, pp. 407–419, 2022.

[659] N. Queralt-Rosinach, R. Kaliyaperumal, C. H. Bernabé, Q. Long, S. A. Joosten, H. J. van der Wijk, E. Flikkenschild, K. Burger, A. Jacobsen,

B. Mons, and M. Roos, "Applying the FAIR principles to data in a hospital: challenges and opportunities in a pandemic," *J. Biomed. Semant.*, vol. 13, no. 1, p. 12, 2022.

[660] C. T. Laurencin and A. McClinton, "The COVID-19 pandemic: a call to action to identify and address racial and ethnic disparities," *Journal of Racial and Ethnic Health Disparities*, vol. 7, pp. 398–402, 2020.

[661] G. E. Ezequiel, A. Jafet, A. Hugo, D. Pedro, M. Ana Maria, O. V. Carola, R. Sofia, S. Odet, T. Teresa, H. Jorge, *et al.*, "The COVID-19 pandemic: a call to action for health systems in Latin America to strengthen quality of care," *International Journal for Quality in Health Care*, vol. 33, no. 1, p. mzaa062, 2021.

[662] Á. Manjarrés, C. Fernández-Aller, M. López-Sánchez, J. A. Rodríguez-Aguilar, and M. S. Castañer, "Artificial intelligence for a fair, just, and equitable world," *IEEE Technology and Society Magazine*, vol. 40, no. 1, pp. 19–24, 2021.

[663] A. Sinha and M. Rathi, "COVID-19 prediction using AI analytics for South Korea," *Appl. Intell.*, vol. 51, no. 12, pp. 8579–8597, 2021.

[664] K. Heo, D. Lee, Y. Seo, and H. Choi, "Searching for digital technologies in containment and mitigation strategies: experience from South Korea COVID-19," *Annals of Global Health*, vol. 86, no. 1, 2020.

[665] H. Chung, H. Ko, W. S. Kang, K. W. Kim, H. Lee, C. Park, H.-O. Song, T.-Y. Choi, J. H. Seo, and J. Lee, "Prediction and feature importance analysis for severity of COVID-19 in South Korea using artificial intelligence: model development and validation," *Journal of Medical Internet Research*, vol. 23, no. 4, p. e27060, 2021.

[666] T. Nam, "How did Korea use technologies to manage the COVID-19 crisis? A country report," *International Review of Public Administration*, vol. 25, no. 4, pp. 225–242, 2020.

[667] J. K. Lee, L. Lin, and H. Kang, "The influence of normative perceptions on the uptake of the COVID-19 TraceTogether digital contact tracing system: Cross-sectional study," *JMIR Public Health and Surveillance*, vol. 7, no. 11, p. e30462, 2021.

[668] H. Stevens and M. B. Haines, "Tracetogether: pandemic response, democracy, and technology," *East Asian Science, Technology and Society: An International Journal*, vol. 14, no. 3, pp. 523–532, 2020.

[669] M. R. Davahli, W. Karwowski, and K. Fiok, "Optimizing COVID-19 vaccine distribution across the United States using deterministic and stochastic recurrent neural networks," *Plos one*, vol. 16, no. 7, p. e0253925, 2021.

[670] J. Jumper, R. Evans, A. Pritzel, T. Green, M. Figurnov, O. Ronneberger, K. Tunyasuvunakool, R. Bates, A. Žídek, A. Potapenko, *et al.*, "Highly accurate protein structure prediction with AlphaFold," *Nature*, vol. 596, no. 7873, pp. 583–589, 2021.

[671] D. V. Gunasekeran, R. M. W. W. Tseng, Y. C. Tham, and T. Y. Wong, "Applications of digital health for public health responses to COVID-19: a systematic scoping review of artificial intelligence, telehealth and related technologies," *npj Digit. Medicine*, vol. 4, 2021.

[672] J. P. Ioannidis, S. Cripps, and M. A. Tanner, "Forecasting for COVID-19 has failed," *International journal of forecasting*, vol. 38, no. 2, pp. 423–438, 2022.

[673] P. Jayanthi, B. K. Rai, and I. Muralikrishna, "The potential of quantum computing in healthcare," in *Technology Road Mapping for Quantum Computing and Engineering*, pp. 81–101, IGI Global, 2022.

[674] G. John, N. S. Sahajpal, A. K. Mondal, S. Ananth, C. Williams, A. Chaubey, A. M. Rojiani, and R. Kolhe, "Next-generation sequencing (NGS) in COVID-19: a tool for SARS-CoV-2 diagnosis, monitoring new strains and phylodynamic modeling in molecular epidemiology," *Current issues in molecular biology*, vol. 43, no. 2, pp. 845–867, 2021.

[675] W. Y. Ng, T.-E. Tan, P. V. Movva, A. H. S. Fang, K.-K. Yeo, D. Ho, F. S. San Foo, Z. Xiao, K. Sun, T. Y. Wong, *et al.*, "Blockchain applications in health care for COVID-19 and beyond: a systematic review," *The Lancet Digital Health*, vol. 3, no. 12, pp. e819–e829, 2021.

[676] S. Juddoo, C. George, P. Duquenoy, and D. Windridge, "Data governance in the health industry: Investigating data quality dimensions within a big data context," *Applied System Innovation*, vol. 1, no. 4, p. 43, 2018.

[677] T. Bernardo, K. E. Sobkowich, R. O. Forrest, L. S. Stewart, M. D'Agostino, E. P. Gutierrez, D. Gillis, *et al.*, "Collaborating in the time of COVID-19: The scope and scale of innovative responses to a global pandemic," *JMIR Public Health and Surveillance*, vol. 7, no. 2, p. e25935, 2021.

[678] P. Dunn and E. Hazzard, "Technology approaches to digital health literacy," *International journal of cardiology*, vol. 293, pp. 294–296, 2019.

[679] D. Wilson, "Daniel Wilson Quotes." https://www.brainyquote.com/quotes/daniel_wilson_741018. Accessed: 2023-12-31.

[680] "Intelligence revolution 1.0: The beginning of the end." https://www.intelligencerevolution.com/

chapter-1-intelligence-revolution-1-0-the-beginning-of
the-end/. Accessed: 2023-12-31.

[681] "Human+digital+artificial   intelligence   ^{tm}   framework."   `https://www.intelligencerevolution.com/`
`human-digital-artificial-intelligence-framework/`.   Accessed: 2023-12-31.

[682] "Technologies of the neolithic era." `https://study.com/academy/`
`lesson/technologies-of-the-neolithicera.html`.   Accessed:
2023-12-31.

[683] "Copper   bronze."   `https://ufl.pb.unizin.org/imos/chapter/`
`copper-bronze/`. Accessed: 2023-12-31.

[684] "Iron   age."   `https://www.traditionalbuilding.com/`
`product-report/iron-age`. Accessed: 2023-12-31.

[685] "Ancient   mesopotamia."   `https://www.getty.edu/art/`
`exhibitions/mesopotamia/`. Accessed: 2023-12-31.

[686] "All   roads   lead   to   rome."   `https:`
`//medium.com/everything-antiquity/`
`all-roads-lead-to-rome-88d5fc4fc011`.   Accessed: 2023-12-
31.

[687] "Telegraph."   `https://www.britannica.com/technology/`
`telegraph`. Accessed: 2023-12-31.

[688] "History   of   cameras."   `https://www.adorama.com/alc/`
`camera-history`. Accessed: 2023-12-31.

[689] "Who   invented   the   automobile."   `https:`
`//www.loc.gov/everyday-mysteries/`
`motor-vehicles-aeronautics-astronautics/item/`
`whoinvented-the-automobile`. Accessed: 2023-12-31.

[690] "Development   of   the   telephone."   `https://www.elon.edu/u/`
`imagining/time-capsule/150-years/back-1870-1940`.   Accessed: 2023-12-31.

[691] "First   flight   of   the   wright   brothers'   1905   flyer."   `https:`
`//centennialofflight.net/essay/Wright\_Bros/1905\`
`_Flyer/WR8.htm`. Accessed: 2023-12-31.

[692] "History   of   computers."   `https://www.toppr.com/guides/`
`computer-aptitude-and-knowledge/basics-of-computers/`
`history-of-computers`. Accessed: 2023-12-31.

[693] "Industrial revolution from industry 1.0 to industry 4.0."
`https://www.desouttertools.com/your-industry/news/503/`
`industrial-revolution-from-industry-1-0-to-industry-4-0.`
Accessed: 2023-12-31.

[694] "Industry 4.0." `https://www.ibm.com/topics/industry-4-0`. Accessed: 2023-12-31.

[695] "Industrial revolution from industry 1.0 to industry 4.0."
`https://www.desouttertools.com/your-industry/news/503/`
`industrial-revolution-from-industry-1-0-to-industry-4-0.`
Accessed: 2023-12-31.

[696] "Jeff bezos on ai." `https://www.cnbc.com/2018/05/11/`
`jeff-bezos-on-ai-robots-wont-take-all-our-jobs.html.`
Accessed: 2023-12-31.

Index

Epilogue

"What is AI?" You can ask a hundred people for a clear definition and each will freeze to think about a palatable answer, but most of them won't possess suitable words for an encyclopedia.

AI is a buzzword that is most often quoted to hide ignorance, if not fear of a dystopic future that will deny humankind the power to decide its destiny. This fear has deep roots; AI can (?) turn into a monster smarter than us, shake us off from the tree of life and discard or replace any one of us, bypassing extenuating factors, to deny us access to credit, if not to our own funds, thus parting us from our birthright to life in a human dimension; this book will be read by scientists or wannabe-such who may object to my affirmation of not-so-PC belief in the G-d of Israel, so anti-scientific. . . The Talmud is a rich wellspring of science, and explained astronomy thousands of years ago in terms that cost Galileo his freedom. And yet, I am not ashamed of standing my ground in a world in which denial of religion is a fig-leaf for thousands of academics who fear for their funding and keep mum about it. Enough of this digression. Caesar's body is still warm. The author of this book is a very dear friend whom I have not yet been privileged to meet in person, but with whom our epistular relation has lasted several years, and I don't feel that I deserve the honor of vouching for this book, since my competence in AI is still growing with my own independent research on intelligent avatars; I am an autodidact inventor and filmmaker, and I am exploring this technology playing by ear, Bach and Mahler in my dreams since early childhood, and Honky-Tonk while awake, with the pedal down to not enrage my bow-tied symphony conductor neighbor, twisting war action videogame notes into my mess of wires and doodads. . .

I have loved reading this book and deepening my knowledge of AI, filling in historical lacks and clearing away long held doubts. I look forward to a few more readings but the time to write these words is now.

To the students who will use this text to follow a course, I offer my blessings of success and congratulations for their choice of a path that will contribute much to human progress; just thirty years ago there were people who thought me fit to be hospitalized when I spoke about the Internet and social media. Now it's AI's turn to be the bugaboo, but in less than thirty years there will be food security and abundance in the whole world, in a sane economic system with AI making true governance a reality, unless it is gamed by greedy agents who will get away with manipulating language into Newspeak and stick us in front of hebetizing videogame consoles.

Good Morning World,
The Singularity is here; enjoy your pizza.
Eliahu Gal-Or Founder: AI-Fixer

Mevo Modiin
December 2023   Israel and Rome, Italy